Lean Implementation
Applications and Hidden Costs

Sustainable Improvements in Environment Safety and Health

Series Editor

Frances Alston

ESH Director
Lawrence Livermore National Laboratory, CA

Lean Implementation
Applications and Hidden Costs

by

Frances Alston

CRC Press
Taylor & Francis Group
Boca Raton London New York

CRC Press is an imprint of the
Taylor & Francis Group, an **informa** business

CRC Press
Taylor & Francis Group
6000 Broken Sound Parkway NW, Suite 300
Boca Raton, FL 33487-2742

International Standard Book Number-13: 978-1-4987-7337-9 (Hardback)
International Standard Book Number-13: 978-1-1387-4797-5 (Paperback)

Library of Congress Cataloging-in-Publication Data

Names: Alston, Frances (Industrial engineer), author.
Title: Lean implementation : applications and hidden costs / Frances Alston.
Description: Boca Raton, FL : CRC Press, 2017. | Series: Sustainable improvements in environment safety and health
Identifiers: LCCN 2016048562 | ISBN 9781498773379 (hardback : alk. paper) | ISBN 9781498773386 (ebook)
Subjects: LCSH: Organizational change--Management. | Organizational effectiveness--Management. | Cost control. | Quality control.
Classification: LCC HD58.8 .A6778 2017 | DDC 658.4/06--dc23
LC record available at https://lccn.loc.gov/2016048562

Visit the Taylor & Francis Web site at
http://www.taylorandfrancis.com

and the CRC Press Web site at
http://www.crcpress.com

Contents

Preface

Lean thinking is a concept that is widely discussed and used on many levels across various business processes. The application of Lean has been visible in many industries, such as automotive, health care, and banking, and in government. Lean in a sentence involves instituting practices that will eliminate nonvalue added steps that will reduce waste and create value while fostering a culture that is supportive of continuous improvement. The value that is created can be beneficial to both the customer and the company. Lean thinking and reengineered processes are helping companies to increase productivity, meet and exceed their customer expectations, and improve their financial goals.

Some practitioners refer to Lean as a process that affords an organization the opportunity to "do more with less." This essentially is true; often Lean process improvement initiatives can result in staff reduction or reallocation and changes in the company compliance posture, impact the ability to successfully implement a succession planning strategy, limit knowledge transfer and employee retention, and the list goes on. These impacts can be costly to a company when balancing the changes that are required to keep these processes optimized and address the people aspects of implementing Lean. The impact and cost associated can be hidden if not addressed during the upfront planning process.

The ability to successfully implement Lean requires that the culture of the organization be open to adapting to changes in the new way that business will be conducted. The lack of a culture that is supportive of change has contributed to the failure of many attempts to implement Lean process improvement initiatives. To successfully implement Lean thinking within a company, a comprehensive strategy must be in place that includes not only the reengineered process, but the people aspects of the process. The strategy should include ways to address issues such as the following:

- Implementation of the technical aspects of the newly improved process
- Employee perception and engagement
- Impact to regulatory aspects such as the environmental safety and health procedures and practices
- Succession planning strategy
- Retraining and technical knowledge retention

- Policies and procedural changes related to implementing the new process
- Considerations for the cultural changes needed to successfully implement the Lean process

This book will address key organizational issues that must be considered and addressed when implementing Lean business practices, offer solutions for many of the challenges, provide a resource that leaders can use in addressing cultural and regulatory issues, and provide a means to address the associated people issues and the challenging task of knowledge retention and succession planning. Vignettes are used to illustrate and provide examples of potential issues and solutions that can be considered for resolving issues as well as identify key references that can be consulted for additional information on key concepts. A case study is included that demonstrates ways to address the technical and people aspects of implementing Lean to ensure project success.

About the Author

Dr. Frances Alston has built a solid career foundation over the past 25 years in leading the development of management and of environment, safety, health and quality (ESH&Q) programs in diverse cultural environments. Throughout her career, she has delivered superior performance in complex, multistakeholder situations and has effectively dealt with challenging safety, operational, programmatic, regulatory, and environmental issues.

She has been effective in facilitating the integration of ESH&Q programs and policies as a core business function while leading a staff of business, scientific, and technical professionals. She is skilled in providing technical expertise in regulatory and compliance arenas as well as in determining necessary and sufficient program requirements to ensure employee and public safety, including environmental stewardship and sustainability. Dr. Alston also has extensive knowledge and experience in assessing programs and cultures to determine areas for improvement and development of strategy for improvement.

She holds a BS degree in industrial hygiene and safety/chemistry, an MS degree in hazardous and waste materials management/environmental engineering, an MSE in systems engineering/engineering management, and a PhD in industrial and systems engineering.

Dr. Alston is a fellow of the American Society for Engineering Management (ASEM) and holds certifications as a Certified Hazardous Materials Manager (CHMM) and a Professional Engineering Manager (PEM). Her research interests include investigating and implementing ways to design work cultures that facilitate trust.

1

Lean Thinking

1.1 Introduction

Many practitioners have argued the sustainability of Lean process improvement in an organization. Some even believed it to be yet another *fad* or *buzzword* that will dissipate with time. Lean process improvement initiatives are not likely to be successful without being championed and supported by a Lean thinking team that includes the leadership team and workers. Lean, simply put, is a way of thinking followed by the right actions that guide the way in which work is performed to improve business across the board in an efficient and streamlined manner. The practices and procedures that support Lean thinking must be ingrained in the culture and the mind of members of the organization. Once Lean becomes ingrained in the minds of the leaders and workers and is seen in their actions, the organization can move forward with purpose and successfully implement a Lean business strategy. Lean thinking is not a temporary state of mind or a way of thinking that can be turned off and on at a whim. It requires a commitment to a physical condition as well as long-term discipline. Lean is focused primarily on providing value for the customer, eliminating waste, and continuously improving all facets of business processes. The most important components that are critical to implementing Lean are presented in Figure 1.1. These components will also be discussed in some form throughout this chapter and this book.

Going Lean can be rigorous and draining on internal resources initially; therefore, careful considerations along with a detailed strategy are necessary to chart the way to success. Before beginning the journey to Lean, it is imperative that top management is supportive and that support is demonstrated through the words they speak and also through their actions. In addition, it is necessary to ensure that the people, physical resources, and policies are in place before moving forward.

Some would say that Lean offers an organization the ability to do more with less; however, a more accurate depiction of Lean is that it offers the ability to optimize processes and outcomes that yield a *win–win* environment for a company, its employees, and its customers. In doing so, often, the

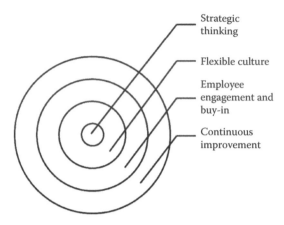

FIGURE 1.1
Lean process components.

greatest negative impact of Lean is experienced by the workers. Many companies have failed in their attempt to implement Lean because the people aspects of Lean were not considered at the up-front planning stage.

1.2 Attributes of a Lean Organization

Before implementing Lean, it is necessary to ensure that the organization is ready and poised for success. It is wise to conduct a review of the organization to determine if the characteristics needed for successful Lean implementation and sustainability are present. There are some fundamental principles that must be kept in the mind of organizational members that are important for success and must be embedded in the business practices. These principles are listed in Table 1.1.

In order to chart the course to implement Lean principles, there are some very fundamental attributes that are important and that must be nurtured. The attributes listed later are not meant to be all-inclusive; however, they do provide a place to begin when attempting to ensure success in implementing the principles of Lean:

Attribute 1: Know your customer.

Attribute 2: Demonstrate respect for people.

Attribute 3: Management should make decision focusing on the long-term value as opposed to *the now*.

Attribute 4: Create a culture that will allow workers to feel free to bring problems to management's attention as soon as they arise.

TABLE 1.1

Overriding Lean Principles

Principles of Lean	Functions
Determine the customer value	Clearly defines the value for products and services that the customer is expecting and targets all nonvalue-added activities for removal from the process. These activities represents system waste
Identify the value stream and then map	The value stream consists of all of the activities that are a part of producing the products or services delivered to the customer. The knowledge gained from this process helps determine what means will be used to deliver what the customer is expecting
Focus on eliminating waste and create flow	Eliminating waste ensures efficient flow of products or services to the customer without interruptions. Waste elimination is also another way to reduce cost
Respond to pull by customer	Understanding what the customer wants and when they expect delivery of products or services. Developing a process to deliver
Pursue perfection to achieve continuous improvement	Continuous improvement in creating flow, identifying, and removing waste

Attribute 5: A continual focus on continuous improvement.

Attribute 6: Maintain a focus on delivering quality continuously.

Attribute 7: Avoid overproduction (use a pull system) where feasible.

Attribute 8: Develop leaders who are inquisitive, walk the talk, and are worker friendly.

Attribute 9: Develop a process that can monitor performance as well as locate improvement initiatives.

Attribute 10: Never stop learning; develop a culture that embraces a learning organization philosophy.

These attributes not only provide the optimal operating environment for Lean, they are also beneficial to other aspects of a business strategy. Each of these attributes will be discussed in a little more detail in Sections 1.2.1 through 1.2.10. These attributes are not listed or discussed in order of preference or priority.

1.2.1 Know Your Customer

It is paramount that you know your customer base keeping in mind that customers can be internal as well as external. Not all efforts performed may add value to all customers and therefore may not be supported by all customers. It is also paramount that the needs of each customer are known and a strategy to deliver quality products and services is in place. A good strategy to consider would be to schedule routine meetings with the customer or ask the customer to complete a customer satisfaction survey at some frequency. It is important to establish and keep a close connection with all customers.

When meeting customers, ensure that the meetings are kept sacred, which means that canceling and rescheduling should be avoided or kept at a minimum. Canceling and rescheduling meetings may give the customer the perception that you do not value their time and business. The meeting times should be used wisely to build relationship and provide insights into what the customer needs are and whether your organization is meeting those needs. If a survey is used, ensure that the survey is designed to provide meaningful data that can be used to determine performance and whether or not the customer expectations are being met. If customer needs and expectations are not being met, it is appropriate to engage in a discussion with the customer on the areas that the company needs to focus on in order to turn performance and the relationship around. Good customer relationship often yields repeated business opportunities.

1.2.2 Respect for People

This attribute can be viewed as one of the most important attributes because it has to do with the people who will be impacted as a result of implementing Lean. Often, we hear leaders communicate that the most important resources in their company are the workers. Without dedicated and skilled workers, a business cannot succeed. Implementation of Lean without careful considerations and involvement of the workers can lead to a loss of trust and respect for the leaders and the company. This can represent a major problem when implementing Lean. Some actions that management should exhibit if it desires to demonstrate respect are listed as follows:

- Persistent and frequent communication
- Praise for a job well done when it occurs
- Listen to ideas and suggestions
- Consult employees on solutions to issues and process improvement suggestions and initiatives
- Be willing to assist when needed
- Saying thanks when appropriate
- Treat people with courtesy and kindness
- Encourage the exchange of opinions and ideas
- Use workers' ideas and input to improve work processes where feasible

When people feel respected, they are likely to respect the management team and the company. The respect of workers will manifest itself in the way they respond to the leadership team and the work they perform on behalf of the organization.

1.2.3 Management Make Decisions Focusing on the Long-Term Value as Opposed to the Now

Decision making is a fundamental responsibility of management. In fact, it is a primary function of a manager. Management effectiveness is based heavily on the quality of the decision they make daily. A decision is systematically defined as selecting a course of action from a set of potential actions to increase the probability of achieving a desired result. Therefore, it is pertinent that decisions are made taking into considerations the long-term strategy to ensure successful and continual implementation of Lean thinking and process improvement. It is vital that management is strategic in decision making, taking into considerations the long-term implications and not seeking only to implement a strategy that will yield only a short-term success. A strategy focused only on the short-term success is not optimal for Lean sustainability.

1.2.4 Create a Culture That Will Allow Workers to Feel Empowered and Free to Bring Problems to Management's Attention As Soon As They Arise

Worker involvement in the workplace is critical in accomplishing work. Not only is worker involvement necessary for work to be performed, it also has a direct impact on productivity and product quality, which are the two important elements that directly impact the financial bottom line of a company. Employees working in an empowered culture tend to be free with providing suggestions and ideas that can improve business processes. The benefits of the workplace culture are discussed in detail in Chapter 3. Also found in Chapter 7 are ways to evaluate and improve culture. Recognizing that the culture of an organization is the key in driving behaviors, managers must spend time in nurturing and fostering workplace cultures that can support Lean thinking and implementation.

1.2.5 A Continual Focus on Continuous Improvement

Continuous improvement of business practices and processes is a way to separate the marginal companies from the good companies. Companies and workers who are happy with the status quo are not likely to achieve greater success and are likely to see their business stifled due to the lack of innovation. This lack of innovation and continuous improvement may cause an organization to lose the ability to compete in their respective markets. This principle is discussed in more detail in Section 1.7.

1.2.6 Maintain a Focus on Quality Continuously

The customer determines quality; therefore, in order to deliver quality to your customers, there is the need to understand the needs and expectations of the customer. Once there is an understanding of the customer quality needs, then a plan should be put into place to ensure that quality is achieved and maintained. Managing quality is crucial for a business because quality products and services can help

- Maintain customer satisfaction and loyalty.
- Reduce the risk of producing faulty and unusable products.
- Build the company's reputation for quality.
- Reduce the cost associated with rework.

There are a host of tools available to assist in evaluating and controlling quality. Many of these tools are listed in Table 1.2.

1.2.7 Avoid Overproduction (Use a Pull System)

This principle is discussed in detail in Section 1.7.

1.2.8 Develop Leaders Who Are Inquisitive, Leaders Who Walk the Talk on a Daily Basis, and Are Worker Friendly

Management by walk around is an effective way of management. In fact, it puts the manager in touch with the workforce and what work is being performed on the shop floor. This form of management is also effective in building relationship with employees. Managers who spend time on the shop floor are typically viewed by workers as being friendly, caring, and

TABLE 1.2

Quality Control Tools

Analysis of Runs	Capability Analysis	C Charts	Cumulative Sum Control (CUSUM) Charts
Expoentially Weighted Moving Average (EWMA) charts	Individual and moving range charts	Lag plots	Levey-Jennings charts
Moving average charts	NP charts	Pareto charts	P charts
Repeatability and Reproducibility (R&R) studies	R chart	S chart	Tolerance intervals
U charts	X-bar charts	X-bar and S charts	X-bar and R charts

approachable. Leaders who are inquisitive often know what questions to ask in order to gage performance and uncover areas that require improvement. They are also viewed by workers as being interested in their work and value their contributions. An inquisitive manager who *walk the talk* is able to get workers to openly provide feedback, embrace change, and participate in important initiatives.

1.2.9 Develop a Process That Can Monitor Performance As Well As Locate Improvement Areas

It is said that we generally do well in the areas that we monitor or are focused on. This is often true because monitoring allows the focus to be directed to a specific process or task. Monitoring can be done through various means; however, before any attempt to monitor performance is embarked on, the performance objectives must be defined. Some steps that can be used to help with process monitoring include the following:

- Define the performance criteria
- Compare the actual performance against the desired performance
- Identify the difference and reason for the differences
- Determine a strategy to correct performance
- Implement and monitor strategy
- Review and analyze data

Many of the tools listed in Table 1.2 can be used to monitor performance as well as to identify process improvements.

1.2.10 Develop a Culture That Embraces a Learning Organization Philosophy

A learning culture has a set of values and practices that encourages individuals to increase knowledge, competence, and performance. In a learning culture, employees are actively engaged in learning, and mistakes are used as a teachable moment for employees and management. In a learning culture, one can expect the following:

- Learning is a daily habit that is demonstrated through behaviors.
- A recognition that mistakes can be an opportunity to learn.
- Mistakes are tolerated with no adverse penalty on employees.

In a culture where mistakes are tolerated, employees feel free to develop the entrepreneurship and ingenuity needed to discover different and more efficient ways of conducting business.

1.3 Why Go Lean?

The aforementioned principles operating within an organization can yield great benefits to facilitate Lean thinking and implementation. One may ask, why is Lean implementation important to business outcomes? The benefits of a Lean organization include the following:

- Lean organizations require less efforts by humans to perform work.
- Lean organizations use less material to create products.
- Lean organizations offer streamlined effective services.
- Lean organizations take less time and efforts to develop their products.
- Lean organizations have a greater focus on the needs and wants of the customer.
- Lean organizations are known for product quality.
- Lean organizations seek ways to continue to improve their products, their way of doing business, and their relationship with the customer.
- Lean organizations have a greater focus on the people aspects of the business.

1.4 Delivering Customer Value

Lean thinking organizations understand the importance of consistently creating and delivering value to the customer. These organizations typically have continuous improvement at the forefront of everything they do. There are different levels of value that the customers may be seeking. Considerations should be given to values that are of importance in economic, business, and personal areas. Figure 1.2 provides a brief summary of what may be included in each value proposition category. Table 1.3 provides some questions that can be used to identify and define each value proposition and help zero in on what the customer may be seeking.

1.5 Defining the Waste Stream

Waste can cost a company a significant amount of revenues that otherwise would be used to enhance other areas of the business. The challenge is to be able to identify the waste that needs to be managed or mitigated. Identifying

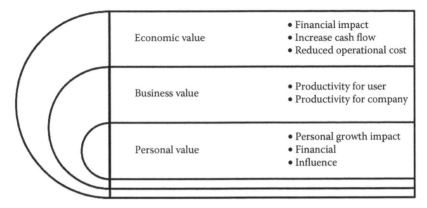

FIGURE 1.2
Customer values proposition.

TABLE 1.3

Value Proposition Identification

Value Proposition	Questions
Economic value	1. What is the return on investment?
	2. Will operational cost be less?
	3. What is the financial impact on the company?
	4. Will revenue or cash flow increase?
Business value	1. What does the investment do for the company?
	2. How can the investment increase productivity?
Personal value	1. Will I get additional compensation?
	2. Will I get promoted?
	3. Will success improve my influence?

and eliminating waste are not generally easy. The first part of the process is to be able to identify what constitutes waste. There are many ways to identify waste. Some of these avenues are listed as follows:

- Ask the worker
- Map out the process
- Review performance data
- Conduct a third-party review of the process by experts
- Benchmark the performance of a similar process

Waste can fall within one or more categories or forms. These forms are listed in Table 1.4 with a brief description of what to look for.

TABLE 1.4

Forms of Waste

Type of Waste	Source
Defect	Products that are not produced based on specifications
Transportation	Moving items from one location to another
Underutilized	Underutilizing skills of worker, workers engaged or not
Human capital	Functioning to their potential
Inventory control	Storing resources such as parts or equipment way in advance of usage
Stalled time	Waiting for instructions or steps to be completed, equipment, and supplies
Overproduction	Producing more than what is needed
Movement	Movement of people, having them perform unnecessary steps and ergonomic issues resulting from awkward movement or performing unnecessary tasks
Talent	Not engaging workers in the business

1.6 Actions That Create Flow

Lean flow is concerned with how people and products move through the system from one step to another or from one station to another. The goal is to optimize the flow of work in a manner in which people and products move quickly and safely throughout the system. Care must be given to avoid jeopardizing quality while optimizing the flow. The optimal goal of an increasing flow is to improve the throughput while improving quality and customer satisfaction. Some actions that can be used to help create a flow are listed as follows for considerations:

1. Focus on the needs of the customer.
2. Determine how work is performed.
3. Identify and remove inefficiencies (waste).
4. Empower workers and encourage them to be engaged in identifying process improvement initiatives.
5. Evaluate how the process is functioning (collect and evaluate the data).

1.7 Produce What Is Used by the Customer and Avoid Overproduction

Overproduction is defined as making products in larger quantities than needed or before they will be used, leading to excessive inventory and the need for a dedicated storage space. One key aspect of producing only what is needed and avoiding overproduction is to use a pull system. In a pull system, products are produced based on fulfilling orders or requests made by the customer producing only what is requested. A pull system is used to maintain small quantities of items needed and replace only what is used. This practice helps avoid overproduction and the need to store large quantities of items for extended periods of time. Reducing overproduction reduces the amount of funds tied up in raw materials, storage space needs, and movement of inventory.

1.8 Seek Perfection to Achieve Continuous Improvement

Continuous improvement in this context refers to ongoing efforts to identify and eliminate problems and to seek and implement practices or procedures to improve quality and efficiency of operation. Continuous improvement is a philosophy that seeks to improve all factors related to transforming a process on an ongoing basis. In order to achieve a culture of continuous improvement, it requires long-term support and investment by top management. Long-term support can include providing training to workers, allocating adequate resources, or continuing to foster a culture of change and employee engagement.

Continuous improvement requires that the entire organization be involved in the process. Employees must be motivated and involved and accept continuous improvement as a way that the company can gain a competitive advantage and become a leader in the business segment in which they compete. Employee engagement and empowerment are key elements in the process of continuous improvement. When workers are active in reporting issues to management and helping management seek solutions to problems, continuous improvement becomes a constant part of the business practices.

1.9 Overview of Lean Tools

In this section, several Lean tools will be highlighted that can be of some use during the transition process. These tools can also be useful in fostering a workplace culture that is needed to sustain process improvement for the long haul. Not all of the tools mentioned will be useful for every change in operation, practice, or procedure. The tools discussed are not represented as an all-inclusive list, recognizing that there are other tools that may achieve the same level of success in implementation.

Recognizing that one of the most important steps in Lean implementation is to gain support of the members of the implementing group, the next step is to select tools that will help identify ways to improve the process and measure performance. Increasingly, organizations are using Six Sigma and the associated tools to identify areas where improvements can increase value and the bottom line for the customer and the company. There are a host of books written on Lean tools that can be used to diagnose, improve, measure, and control processes.

1.9.1 Value Stream Mapping

There are many tools that can be used to identify, diagnose, and implement Lean process improvement initiatives. However, the beginning of the evaluation process should start with mapping the process. Process or value stream mapping is effective in providing a pictorial chronological view of a project, task, product, or process. It provides a view that allows the flow of work and the process activities to become transparent, so that improvements in the process can be adjusted to remove waste and increase efficiency. This view provides a comprehensive look and a unique opportunity to analyze each step or activity to improve efficiencies and eliminate waste.

Mapping is critical because during the mapping, the *as is* design is determined and the *to be* design is defined. The value stream consists of all activities involved in the process, task, or group involved in producing the output. The identification process is most effective and comprehensive when performed in a team setting. It is also critical to ensure that the right team members are assembled to ensure optimal efficiency. At the conclusion of the process, it should be clear what practices or steps in the process are nonvalue added and can be discarded. Several tools that can be used in identifying process issues and improving efficiency in operation are included in Table 1.5. Many of these tools can also aid in monitoring system performance.

1.9.2 Six Sigma and Lean

The goals of Six Sigma and Lean are in alignment because both are designed to improve process efficiency, eliminate waste, improve quality, and create efficiency in systems. This is accomplished by streamlining and improving processes across the business. Six Sigma was originally designed to be used in manufacturing. However, it was quickly discovered that Six Sigma has broad applicability throughout all aspects of a business process. Like Six Sigma, Lean is a tool that when used effectively can help to streamline business processes. The Six Sigma tool kit consists of a host of tools to help analyze performance, identify improvement initiatives, and eliminate waste in resources. The methodology of Six Sigma is typically implemented in five

TABLE 1.5

Lean Tools

Tool	Description	Value Proposition
Andon	A visual system that displays the status of the process to include providing an alert when assistance is needed and provides empowerment to workers to stop production when they determine it to be appropriate	A real-time communication tool that provides immediate notification of issues that allows management to immediately address problems as they occur
Bottleneck analysis	Used to identify the part of the process that limits and hinders complete efficiency	Improves throughput by removing and strengthening the weakest part of the process
Cellular manufacturing	Process to simplify workflow and focusing on a single product or a specific batch	Used to organize a small team of people to focus on a single or small batch of products completing the entire product prior to leaving the work cell
Continuous flow	Process designed to allow work to flow smoothly through production with ease	Eliminates several forms of waste such as time, inventory, and material
Gemba	A philosophy of encouraging field presence of everyone from senior management to the worker level	Promotes a top-down understanding of the actual process and potential issues
Heijunka	A scheduling technique using sequencing and producing products in smaller batches	Reduces lead time and inventory because batches are smaller
Hoshi Kanri (policy implementation)	Alignment of the company's goals and the work that is being performed	Ensures that the strategic plan is consistently implemented throughout the company
Jidoka (automation)	Design the process to take advantage of partial automation. The process should be flexible enough to allow automation to cease if defects or issues are detected	Less expensive than complete automation. Potently to reduce labor cost because workers can monitor multiple stations
Just in time (JIT)	Purchase (pull) parts based on known demand	Reduces inventories and space requirements and improves cash flow
Kaizen	A strategy proactively used to continuously improve process, procedures, or technologies using inputs from employees at all levels of the company	Uses the team approach and combines the talents within the company to facilitate improvement. Works well with standardized work

(Continued)

TABLE 1.5 (*Continued*)

Lean Tools

Tool	Description	Value Proposition
Kanban	A method for regulating the flow of goods internally and externally using a system of automatic replenishment when more supply is needed	Eliminates the need to maintain inventory as well as the act of inventorying goods
Key performance indicator (KPI)	Metrics that are designed to track the progress of critical goals. These metrics can guide behavior because they are typically reviewed frequently by management	Key performance indicators are effective because they should be aligned with senior management strategic goals and can drive the behavior of those who are critical to achieving results
Mistake proofing (Poka-yoke)	Combines Ishikawa charting and Pareto analysis to analyze and identify process-related issues	Helps to identify process and procedural issues to improve workflow
Plan, do, check, act (PDCA)	The methodology involve establishing the plan, implementing the plan, verify that the results received are the expected results, and review and assess	Applies a strategic approach to improving process, practices, and procedures
Quick changeover	Converting an equipment from servicing one product to another in a short time period	Short changeover times can be effective in reducing batch sizes and facilitating just in time production and delivery
Root cause analysis	Problem-solving method that focuses on discovering the underlying issue as opposed to leaping to the solution	Effective in ensuring that problems are eliminated by instituting the appropriate corrective action or actions that caused the problem
Six Sigma	A rigorous methodology that uses statistics to improve process and work practices	Improvement in quality and performance in practices and process
Statistical process control (SPC)	Uses a variety of analytical and measurement technique to determine if the process is in control or operating as intended	Using statistics to improve quality and process capability
Takt time	The average rate that a product or action must be transacted based on customer requirements. It is the pace of production that will meet customer needs	A simple and consistent method and pace of production can increase process efficiency

(*Continued*)

TABLE 1.5 (*Continued*)

Lean Tools

Tool	Description	Value Proposition
Task simplification	Developing simple techniques and steps to accomplish tasks such as using engineered tools to help streamline and simplify work	Reduces wasted time and steps
Task standardization	Organizes work and trains workers to complete task the same way each time using the same steps and procedures	Reduces variability in work method, process time, and product quality
Total quality management (TQM)	Maximizes the use of teams to solve problems and improve process capability to ensure that process is kept in control	Facilitates quality improvement by preventing or reducing defects from occurring
Total productive maintenance (TPM)	Facilitates a maintenance program that is strategic in combining predictive and preventive maintenance to enhance problem solving and product quality	Improves equipment run time as well as product quality
Value stream mapping	Shows the flow of materials and information that allow the ability to categorized out come into three categories: value enabling, value adding, and nonvalue adding	Focuses on identifying and eliminating nonvalue adding steps and practices in order to gain efficiency
Work standardization	Establishes uniformity of working conditions (tools, equipment, procedures, materials, etc.), provides consistency in performing tasks and implementing business processes	Focuses on uniformity in quality and performance
Work balancing	Scheduling and dividing work to minimize idle time for people and equipment	Using simple techniques such as workflow or bar charts to help assign tasks to people and equipment
5s	Focuses on housekeeping through optimal organization (sort, set in order, shine, standardize, and sustain)	Organization of the workplace in the safest, efficient, and most effective manner
5 whys	A process of asking the question *why* several times to get to the root of a problem	Effective in identifying cause and effects

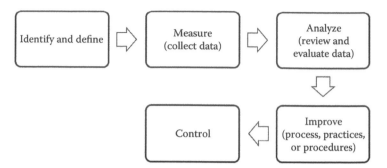

FIGURE 1.3
Six Sigma phases.

stages and relies on the collaboration of a team of professionals with varying knowledge levels and skills. These stages are shown in Figure 1.3.

There are three key factors that are required for successful implementation of Lean and Six Sigma that are consistent and can make the difference between success and failure. These factors include the following:

1. Senior management's involvement
2. Employee engagement
3. A culture that facilitates embracement of change

Six Sigma when used in conjunction with Lean concepts is a proven process improvement methodology that has been used by many companies successfully to combine the benefits of both Lean techniques and Six Sigma to help streamline and continuously improve operations, increase value, and reduce waste.

1.10 Staying Lean

Once a process has been streamlined and is operable, maintaining the Lean status may not be easy. Many attempts to implement Lean have failed for various reasons. Staying Lean requires a commitment to the process, a willingness to think Lean, tools to evaluate and assess the process, and a strategy of continuous improvement (Figure 1.4). Implementing Lean requires a specific way of thinking that is embedded in the strategy of the organization and is supported by the leadership team as well as the workers. The culture of organizations that are successful in facilitating Lean is one that is flexible and is able to embrace change.

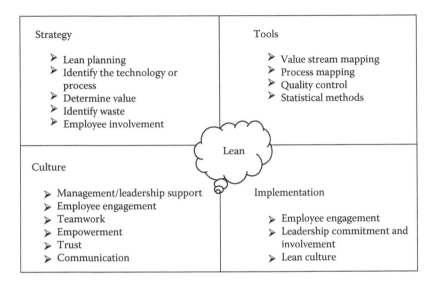

FIGURE 1.4
Staying Lean concepts.

Some of the reasons for Lean failures are as follows:

- Fear of the unknown
- An unwillingness to change
- Systems to implement Lean are not in place
- Culture not conducive to change
- Lack of management support
- Lack of workers' support
- Workers are not engaged in the process
- Lack of trust
- Lack of customer focus
- The organization has a track record of failure when implementing new process and practices
- Lack of a sustainable strategy

1.11 Summary

A Lean thinking leadership team is at the core of successfully implementing Lean in a group, an organization, or a company. Lean thinking represents a strategy accompanied by a way of thinking that is focused on the customer and continuous improvement in delivering quality products and services.

Lean success depends on the culture of the organization, employee engagement, management support, and a strategy that can be implemented and sustained. Lean offers the ability to create more value for the customers with fewer resources, waste, and obstructions. There are many tools available to assist in evaluating a process, technology, or procedure that will facilitate the implementation. These tools provide the means to evaluate processes and determine improvement needs as well as help monitor progress to validate whether a process is in control or out of control based on the defined criteria. Lean requires involvement and the commitment of the leadership team and engagement from the workers in order to be successfully implemented and sustained.

2

Lean Thinking Leadership Team

2.1 Introduction

Good leadership is essential for businesses to conduct work and successfully compete in the market place. Lean thinking leadership is necessary to develop, implement, and sustain Lean policies, practices, and procedures in organizations. A skilled leader is someone who possesses the ability to bring people together and guide them toward achieving a common goal. Anyone can give instructions to others on what to do, but effective leadership requires much more than the ability to assign and track tasks to completion. It requires the ability to think strategically and inspire others to follow. The abilities of a leader can determine if the organization will succeed or fail.

The leader is the conduit and the glue that holds an organization together and charts the course of its members. Leaders have responsibilities such as setting vision and goals for the organization, motivating people, and guiding employees through the work process. A Lean thinking leader is essential in implementing Lean concepts in an organization, getting buy-in from workers, and facilitating employee engagement and a culture that supports and encourages Lean thinking and Lean practices.

A Lean thinking leadership team has a strategy of continuous improvement that is ingrained in the way they conduct business daily. The strategy includes a constant focus on detecting ways of eliminating waste and providing more value for its customers.

Lean thinking has been successfully used to improve processes in various industries in areas such as the health care, technology improvement, and improvement in human resources process and practices. Implementing Lean is a viable way for improving business outcomes across the globe.

2.2 Leadership Responsibilities in Organizations

Organizations are built by leaders and the way an organization functions is based on the leader's style experiences, and knowledge. This philosophy is important to consider when hiring leaders for various roles and functions. There are many important roles that leaders are expected to excel in as they lead their staff and the organization. For example, leaders are expected to have the ability to negotiate, communicate, motivate, and influence others to follow.

The critical roles that leaders fulfill in organizations are shown in Figure 2.1. The effectiveness of a leader in fulfilling these four roles determines the effectiveness of the leader and the level of success that is achievable for that organization. Lean thinking must be an integral part of each leadership tenet to be effective and sustaining. The four roles are further defined in Sections 2.2.1 through 2.2.4.

2.2.1 Vision Tenet

Where do you want the organization to go? A vision is most effective when it is brief and clear enough to encourage buy-in and concrete enough for people to see, understand, and willing to embrace and follow. A vision statement is most effective when it inspires and motivates members of the organization. The vision will not become reality if it is not shared. In organizations where the leadership is without a vision, the organization (people, process, and mission) is unable to move forward and prosper. A good vision statement has the following characteristics at a minimum:

1. Meaningful to the workers and the leadership team
2. Establishes the standard of excellence for the organization

FIGURE 2.1
Basic Tenets of leadership.

3. Links the focus of the present with the focus for the future
4. Implementable
5. Inspire commitment
6. Energizes people
7. Facilitates buy-in (shared vision)

2.2.2 People Tenet

People in an organization look to the leader to chart the course. People will follow a leader they trust and respect. They also expect that the leader will have their best interest at heart. An important aspect of being an effective leader is the ability to develop people and prepare them to consistently and effectively contribute to the vision of the organization. Leaders who focus primarily on their own development and career will not be focused on the needs of the workers.

2.2.3 Strategy Tenet

A strategy is the creation of a valuable position for the organization and development of a road map to implement the activities to achieve the end results. As with Lean thinking leadership, strategic leaders leverage their ability to think critically, anticipate issues before they occur, interpret data and information from various sources, learn from mistakes, and make decisions that are in the best interest of the organization, stakeholders, employees, and customers.

2.2.4 Decision Tenet

Making decisions is expected of leaders. It has been echoed time and again that great leaders are expected to make good decisions. This is primarily due to the understanding and the ability of these leaders to balance emotions with reason while taking into considerations the facts presented through information such as data. All decisions are made in the best interest of the organization, employees, stakeholders, and customers. The primary ways in which the leaders make decisions are as follows:

1. *Collaborative*: Decisions are made based on information and feedback from the team. In such cases, the leader makes the final decision by dissecting the information provided.
2. *Consensus*: Through process of voting in which the leader's decision is the result of the vote.
3. *Command*: The leaders make the decision based on their knowledge and perceptions.

2.3 Followership and Leadership Roles

Followership is a straightforward, yet simple concept that many leaders do not fully understand. Simply defined, followership encompasses one's ability to accept direction from another person, to support a concept, strategy, project, or program that was derived by someone else, and to deliver on what is expected of an individual. Focusing on followership can provide an insight into how leaders can become and remain effective. If one can gain an understanding of why people follow leaders, they can impact the development of followers who have the potential to become effective leaders. People do not just follow anyone because they are requesting them to follow. They must have a reason that they have bought into in order for them to follow.

When employees trust their leader, they are willing to follow. There are some key attributes that must be present that forms the basis for the trust between employees and their management team. These trust attributes are shown in Figure 2.2 and defined in Table 2.1. Trusted leaders are able to excite workers to become followers (Figure 2.2).

It has been established that people will follow those they are connected to and trust; without trust, there can be no followership. Therefore, trust is a

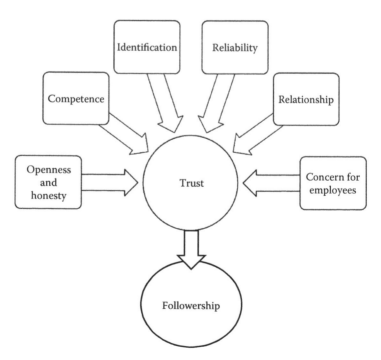

FIGURE 2.2
Trust impact on followership.

TABLE 2.1

Trust Attributes Defined

Trust Attributes	Definition
Openness and Honesty	Leaders can demonstrate these qualities through the things they say the things they do. The information disclosed is complete, accurate, and timely
Competence	Refers to the leader's ability to function effectively as a leader. Does not refer to the leader's technical abilities in the function he or she is leading
Identification	Refers to the extent to which groups hold in common goals, norms, values, and beliefs
Reliability	Refers to the leader's reliability in making appropriate decisions and providing support to employees
Relationship	Refers to the employee's ability to communicate and feel a connection with the leader and coworkers
Concern for employees	Being sensitive to and understanding the needs of employees. Demonstrating concern deals with a feeling of caring, empathy, tolerance, and concern for safety and well-being of the worker

key attribute in establishing relationships between people that make them feel connected and inspire them to become followers.

2.4 Qualities of a Good Leader

Some leaders start out as poor or marginal leaders. Good leaders emerge after experience and additional training. However, when selecting leaders, certain skills should be present that could indicate whether they have the ability to become a good leader. Aspiring leaders should show some resemblances of the following skills before being selected to take on leadership roles:

1. *Engaged in the business*: Employees who are engaged generally have knowledge of how the business operates and are visible to coworkers and members of the leadership team. They will also often freely share their knowledge and offer new ideas and solutions to problems.
2. *Keep commitments*: Employees who have a demonstrated track record of keeping commitments are often trusted and can have the ability to influence others in the work place.
3. *Good listener*: Able to listen and provide open and honest feedback after hearing the entire story while demonstrating respects for the input and ideas of others.

4. *Effective communicator*: Employees who are able to communicate are naturally viewed by coworkers as skilled in their ability to lead. They are able to get their thoughts across effectively and are generally able to convince others to buy-in to their thoughts.

5. *Able to build and maintain relationship with workers and management at all levels*: They are able to build collaborative relationship with coworkers. These relationships often lead to trust among workers.

6. *Able to make decisions and see them through*: Does not waver in decision making and is able to follow the decision to the end.

7. *An effective team member and leader*: Able to serve effectively in the role of a leader as well as a follower.

8. *Willing and able to delegate*: Not threatened by allowing others to take on the role of a leader or complete a project started by them. Delegation demonstrates that one is willing to extend trust to others and therefore receives the trust of others.

9. *Exhibit trust worthy behaviors*: Encourage followers to follow and serve as an example to others in the organization.

A good leader possesses many good qualities that are important in excelling in his or her role as a leader. Some of these qualities are obvious and some are not so obvious. We will touch upon many of those characteristics in this section. These qualities are just as important for Lean thinking leaders who are constantly seeking ways to improve their processes and inspire others to follow and become Lean thinkers.

2.4.1 A Good Leader Is Confident

A confident leader inspires confidence in his or her followers. Confidence is an important attribute of a leader if the expectation is to inspire others to buy into a vision and act. Leaders who are able to convey confidence in decision making and achieving objectives inspires team members to put forth their best efforts. Confident leaders provide the type of energy that employees can tap into and gain inspiration that facilitates the desire to follow their leader. It also encourages innovation in decision making and problem solving, and the ability to accept and handle changes with ease.

2.4.2 A Good Leader Acts with Purpose

Because people look to leaders for solutions and comfort in time of uncertainty, leaders must function in a purposeful manner. Leaders must reassure and demonstrate confidence with a positive demeanor. A lack of direction and purpose makes a leader appear to be indecisive and not worthy of following. This occurs because the follower is unable to determine if the path is useful and well thought out in order to achieve success.

Purposeful leadership is the key to relieving stress among workers during times of uncertainty and change.

2.4.3 A Good Leader Demonstrates Exemplary Character

Good leaders are known for their exemplary character. Character is a quality that distinguishes a great leader from a bad or marginal leader. These leaders have integrity, courage, and passion for the things they do and the people they serve. The character of the leader is visible in the decisions he or she makes, in the words he or she speaks, as well as in his or her actions. A leader needs to be trusted and known for his or her trustworthiness in order to be viewed as effective. Leaders with good character are able to achieve results that translate into high organizational performance.

2.4.4 A Good Leader Is Enthusiastic

People will respond to a person who openly demonstrates passion and dedication for his or her role as a leader. Leaders need to be able to be a source of inspiration and motivation for followers and must be viewed as a viable part of the team working toward the goal. A good leader is enthusiastic about his or her work and his or her role as a leader. This enthusiasm shows in the way he or she manages the day-to-day business as well as in his or her interactions with colleagues, stakeholders, customers, and workers.

2.4.5 A Good Leader Is Focused

A good leader is able to think strategically and analytically while remaining focused on the goal of the organization. Not only does a focused leader view a situation as a whole, but is able to break it down into smaller parts for closer inspection. When speaking of the leader's ability, reference is being made to how the leader approaches all leadership functions. It does not mean that the leader is so focused until he or she allows himself or herself to become myopic and lose touch of other issues. Although focused, the leader is able to demonstrate flexibility when the environmental conditions and the situations that need to be addressed require him or her to change his or her course.

2.4.6 A Good Leader Is Able to Build and Retain Good Relationships

More and more today, businesses are based on the relationships that have been formed among the customer, clients, workers, colleagues, and stakeholders. The ability to build and retain lasting relationship is the key to attracting and retaining customers, clients, and workers. A leader who is capable of building good relationships is able to facilitate trust among people and improve the workplace environment. Good customer relations can lead to repeat businesses and being recommended to other potential clients by current customers.

2.4.7 A Good Leader Is Committed to Continuous Improvement and Excellence

A good leader is committed to getting things done right the first time. The good leader not only sets and maintains high standards but is also proactive in continuously raising the bar in order to achieve excellence. There are so many tools available that can assist leaders and organizations in continuously improving how they operate and serve their clients. Lean thinking leaders excel in selecting the right Lean tools to create value and continuously improve their product, services, and the performance of the organization.

2.5 Lean Thinking Behaviors

Human behavior has been widely studied as it is recognized to have an impact on the way we form and maintain relationships, communicate, deal with people, and the way we think of an individual. Human behavior can fall within two categories: productive or destructive. Productive behaviors are those behaviors that facilitate and yield benefits in relationship, leadership, and followership. Lean thinking leaders exhibit and model the following behaviors every day as they conduct business and serve their customers:

1. They know the importance of the services they provide to their customer through
   • Understanding the needs and wants of the customer.
   • Knowing the importance of the services rendered to the customer.
   • Continuing to seek ways to improve services provided.
2. They demonstrate a mindset of continuous improvement
   • Challenge the status quo and empowering workers to challenge the status quo.
   • Taking advantage of improvement opportunities.
   • Understanding and knowing when the customer needs changes.
3. They invest in employees
   • Guide problem solving.
   • Investing in knowledge gaining and sharing.
4. They focus on and have a knack for obtaining results
   • Known for achieving results.
   • Empowers and inspire others to achieve results.
   • Use resources effectively and efficiently.

5. Facilitates a culture to create and sustain continuous improvement
    - Facilitates a learning culture–using each failure as a teachable moment for self and workers.
    - Modeling and encouraging others to adopt and embrace Lean.

2.6 Walking the Talk

It is easy for some people to say the right thing at the right time. However, managers and leaders must be deliberate in the things they say and ensure that their words match the things they do. Why is walking the talk an important attribute for leaders? It is important because the action of a leader directly impacts the actions of his or her followers. More and more today, managers are encouraged to do what they say when leading as a necessary means to achieve success. In order for managers to be taken seriously, they must be first partakers of following polices, rules, and instructions if they expect others to follow. Leaders are not exempt from following rules and procedures that are set forth within the organization.

2.7 Leadership Styles

Although we often use the terms managers and leaders interchangeably, they are two different roles. Both roles are important and play a vital part in the success of accomplishing work and rendering a company or business successful in its goals and mission. Managers are viewed as facilitators and are responsible for ensuring that workers have all of the tools needed to accomplish work successfully. On the other hand, leaders can be anyone with a talent and strength to get others to perform a specific activity or task and follow their instruction or advice. Leaders have vision and have the ability to recognize when change is needed and have the ability to energize the people and the organization. They also know and recognize what needs to be done and when it should be done to optimize event timing.

In this section, some of the predominant leadership styles will be presented. It is important to know your leadership style when implementing Lean and know when there needs to be a change in style to deal with environmental changes. Many leaders have a primary leadership style and often change styles based on the situation or issue presented. Thus, it is necessary to understand the various styles and how they can potentially support or hinder the leadership team and the organization (Table 2.2).

TABLE 2.2

Leadership Styles

Leadership Style	Description	Advantages	Disadvantages
Autocratic (Authoritative)	Provides little to no supervision for employees. Leader dictates and controls all decisions made	Has a great influence on shaping the culture of an organization	Little feedback is sought from team members; there is a sense that team members or subordinates are not trusted
Delegative (Laissez-Faire)	Decisions are made by the group. Little guidance from the leader. Managers make decisions without impact from others. Possess total authority	Leader empowers others to act, help build consensus and commitment, encourages ideas and creativity, and recognizes the knowledge of others. Highly skilled and experienced workers who require little supervision is optional for this leadership style	Continuous improvement may suffer waiting for the group consensus. Decisions are not made quickly, inexperienced employees may suffer and feel like they are unable to contribute, and the leader may be viewed as indecisive or unable to lead. Not optimal for developing employees
Democratic (Participative)	Leadership style facilitates conversation, encouraging people to share their ideas, and use the information gained to make decisions. The leader consults and involves the group in decision making	Great for work performed in flexible environment. The leader still is responsible for guiding decisions and override if needed	The decision making process may be lengthy
Situational	Leader changes style to meet the needs of others and the organization, four styles of leadership: directing, coaching, supporting, and delegating	Stresses flexibility and simplicity in task execution	Significant influence on decision making based on the current environment
Transactional	A leader uses reward and punishment as a form of motivation. Defines clearly the role of the leader and expectations of followers, assumes that subordinates are only motivated by rewards	Rules are clearly known and understood, effective in motivating and encouraging subordinates and obtaining productivity	Failure to deliver the expected output often results in negative consequences such as reduction of pay, absence of bonuses, or dismissal. Can create stress on workers and fuel fear

(Continued)

TABLE 2.2 (Continued)

Leadership Styles

Leadership Style	Description	Advantages	Disadvantages
Transformational	Tend to exhibit emotional intelligence, tends to be energetic, and passionate about what they do	Committed to helping the organization achieve its goal and help organization members reach their potential	May struggle at times with organizing divergent details, reliance on emotion and passion, thereby overlooking reality
Coaching	Leaders define roles and tasks of followers. Although decisions are made by the leader, input and suggestions are sought	Very successful in improving results. Two-way communication is used continuously and provides a positive work environment	Not used enough because of the investment time. Management may not believe that it has the time to invest
Affiliative	Promotes harmony and effective in conflict resolution and successful in building teams that are effective in accomplishing tasks	Workers are engaged and feel welcomed, valued, and more satisfied about work	Leader does not truly lead. Team members determine the goals and objectives. Leader may have a hard time dealing with conflict
Servant	The servant–leader philosophy is to serve the people in doing so, puts the needs of others first, and engage in people development	Effective in gaining respect, team building, and trust	Employees may view the manager as catering and may not take the leader's authority seriously
Visionary	The leader has an inspiring vision, and is able to help others to see how they can contribute to this vision	Able to inspire the followers to move together toward a shared vision. Promotes innovation, creativity, learning, and relationships	Easy for the vision to be lost in complex and difficult times

(Continued)

TABLE 2.2 (*Continued*)

Leadership Styles

Leadership Style	Description	Advantages	Disadvantages
Charismatic	Leader has the personality that inspires and motivates. His enthusiasm is catching, and team members may gladly follow such a leader to achieve objectives that they could not imagine reaching without them	Able to articulate in a way that is convincing and trusting while arousing emotions in the followers to gain buy-in for the vision	In the absence of the leader, the team may fall apart and lose direction. Charisma may wear thin with time and become less effective
Bureaucratic	Bureaucratic leaders rely heavily on rules, regulations, and clearly defined positions	Expectations and chain of command are clear. Leadership is impersonal and focuses on performance, not the worker	Workers typically develop low morale. Employees are not empowered or engaged in work decisions. Impersonal leadership, very little attention to worker, and focus on strict adherence to rules
Task oriented	Assigns roles, tasks, and responsibilities to team members to include a delivery deadline and expects deadlines to be met	Provides clear direction and expectation on task completion	Team members may perceive that their well-being and abilities are being overlooked because of the focus on efficiency

2.8 The Influential Leader

A leader cannot lead without the ability to influence, because influencing others is how leaders lead. Specifically, leaders lead through their ability to influence followers to adopt and support their vision and to follow in their footsteps. A leader who does not have the ability to influence does not have the ability to be successful in the role of a leader because a large aspect of the job of a leader is based on his or her ability to influence others. Without influential power, a leader is ineffective in accomplishing the goals of the company. To effectively influence others, the ability to develop and grow relationships is essential. Therefore, it can be said that another important role of a leader is to focus on building meaningful and effective relationships with colleagues, subordinates, customers, and followers. Some ways in which influence can be formed are listed as follows:

- Influence is facilitated through connection between people.
- Influence is facilitated through relationships.
- Influence is facilitated through trust among leaders and subordinates.
- Influence is facilitated through display of competence.

Leaders must have the ability to influence others to follow. Because the sphere of influence and the process of influence are different for each leader, for a leader to be effective, he or she should be aware of his or her environmental conditions and the characteristics of the people whom they are trying to influence.

2.9 The Accountable Leader

The ability to execute work and deliver results to the customer is tied to accountability that can dictate the attitudes, practices, and systems that are in place in an organization. Accountability is an important principle that defines how we make commitments to each other and how we react when things go wrong. Greater accountability eliminates the time spent in unproductive behavior such as blaming that produces wasted effort and confusing distractions. A leader who absolves himself or herself of responsibilities is unable to effectively lead others and sets a tone for others holding themselves accountable. On the other hand, a leader who holds himself or herself accountable will see others following pursuit and holding themselves accountable. An accountable organization can expect to have employees who are engaged, team alignment, and trust among workers and leaders.

Accountable leaders hold themselves accountable for their actions and decisions and are able to hold others accountable for the same. Discussions on accountability are constantly sparking discussions on how one would know when it exists within reasonable bounds. With the focus on having a safety conscious work environment, worker involvement, and ensuring that the work environment is appropriate for workers to feel safe and free to communicate, some managers are not always sure when accountability may cross the line into unfair targeting of workers.

2.10 The Lean Thinking Leadership Team

Lean thinking is a mind-set that must be constantly embedded in the thoughts of leaders. Lean thinking leaders tend to have exceptional leadership qualities. These leaders demand respect by nature of their ability to be strategic and inspire others to adopt the goals of Lean thinking within the organization. Customer focus and continuous improvement are at the forefront of their minds and are important aspects of development and implementation of their business strategies.

Lean thinking leaders focus on and consider the following with each decision they make to ensure they deliver value to their customers:

- Define and specify the value proposition
- Identify and pursue the needs of the customers
- Focus on eliminating waste
- Seek to continuously improve
- Seek perfection although it may not be fully realized

Lean thinking leaders focus on and consider the following with each decision they make to ensure they develop the culture needed to facilitate Lean thinking:

- Employee engagement
- Facilitate employee commitment to the goals of the organization
- Develop and implement policies consistently
- Develop a culture of trust
- Communicate honestly and completely

Lean thinking leaders focus on and consider the following to facilitate the development of a Lean thinking team:

- Involve workers in identifying and resolving problems
- Stress continuous improvement
- Invest in employee development and growth

2.11 Summary

It is clear that many organizations are focusing on restructuring their business practices to become more efficient and effective in the services they deliver. In order to accomplish this change in strategy, the appropriate team of leaders must be assembled with the appropriate skills to carry out the strategy. Practitioners and scholars have consistently stated that there is evidence to support that Lean thinking is an effective approach to improving quality while reducing cost and time. Going Lean takes strategy, time, and a strategic thinking team of leaders who are willing to go the course.

Lean thinking is gaining popularity across the globe as companies make concentrated attempts to improve their bottom-line profits and provide their customers exceptional products and support. In the same vein, many attempts to implement Lean have failed some due to the actions of the leadership team and their inability to develop and maintain the right attitude and behavior for success. Lean has no chance of success without the leadership team. Therefore, it is important that the leadership team demonstrates the proper leadership characteristics necessary to lead the organization toward sustainable success in implementation and continual improvement.

3

Lean Culture

3.1 Introduction

More and more corporate leaders are focusing on the impact of culture on business outcomes. The culture of an organization is a very important aspect of how employees think, act, and react to issues, the management team, policies, and producers. There are several elements of culture that must be recognized and considered when making decisions that can impact the organization. Culture can also have an impact on other aspects of an organization such as the following:

- The way rewards are distributed to organization members
- The way people are treated by management
- The process used for promotion
- Implementation of new technology
- Lean process improvement
- Building and retaining trust among organization members
- Knowledge and employee retention

Strong cultures encourage members to adapt to changing environment, explore different ways of accomplishing work, increase desire to work as a team, and increase trust among members. Culture is one of the most important aspects of a company that can set the stage for success or failure when it comes to implementation of process improvement initiatives. This chapter will cover the attributes of a culture that will be optimal and supportive for implementing Lean process improvement initiatives.

3.2 What Is a Lean Culture?

Organizational culture is a system of elements consisting of practices, behaviors, symbols, language, assumptions, and perceptions shared by its members. A Lean culture encompasses all of the attributes of a culture that one would expect in every organization. In a Lean culture, there is also the inclusion of the persistent business mind-set that is demonstrated in every action and decision made by management and workers throughout the organization on a daily basis. The mind-set is a continual focus on decisions and actions that can lead to improvement in process efficiency, product and service quality, worker involvement and growth, and providing exceptional customer service. These attributes have a strong influence on how people act, react, and perform work. Lean thinking is the driving force behind having a Lean culture that can continuously handle process improvement as well as sustain it over time. Attributes of a Lean culture are shown in Figure 3.1. These attributes will be discussed in brief in this chapter with the exception of Lean thinking. This attribute is discussed in more detail in Sections 3.2.1 through 3.2.6.

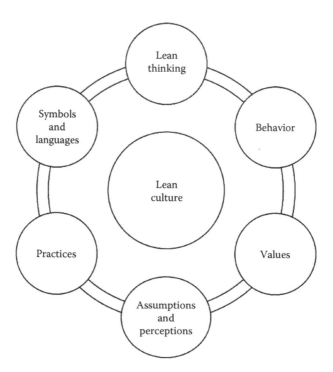

FIGURE 3.1
Lean culture attributes.

Throughout this chapter, the term Lean culture will be mentioned many times. Lean culture is defined as a culture that has all of the elements and attributes required to implement and sustain Lean process improvement initiatives. Characteristics of a Lean culture include the following:

- The organization invests the time needed to build relationship with employees, suppliers, and customers.
- Leaders have a vision and a strategy that are implemented and reevaluated routinely.
- Improvement initiatives are recommended and initiated by people throughout the organization and not only from the leadership team.
- Leaders are in constant touch with customers and organization members.
- Lean tools are frequently used to improve process and measure performance.
- There is a constant push to standardize and streamline processes and to eliminate nonvalue-added steps and activities.
- Workers embrace learning and knowledge sharing.
- Workers are not afraid to try new ways of doing things.
- There is a constant attempt to eliminate waste throughout the process.
- Continuous improvement is a way of life and is embraced.

3.2.1 Practices

The practices that one would expect in a Lean organization demonstrate its capability for success and the ability to be flexible when changes are implemented. When referring to practices, reference is being made to habits that are consistently performed and seen by the employee and management team within the organization. These habits form the basis for consistent actions and the way work is performed within the organization. A habit of acceptance gained through practice can be instrumental in helping employees to remain focused on quality, the needs of the customer, and to accept organizational changes with ease.

3.2.2 Behavior

It has been demonstrated by many theorists and practitioners that culture has a distinct influence on the behavior of organizational members. Cultural behavior in the workplace describes the attitudes, reactions, and mind-set ingrained in its members. The practices and habits that are a part of the organization will drive the behavior of the workers (Figure 3.2).

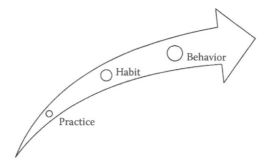

FIGURE 3.2
Behavioral path.

The behavior influenced by culture can have a distinct impact on implementing changes within the organization. More specifically, behavior will impact the way employees approach their daily roles and responsibilities, and adapt to changes as well as their ability to expeditiously support change.

3.2.3 Values

Values can be viewed as a guiding principle of a culture that forms the basis for how members set priorities and carry out the associated actions. It is also an important element that helps to shape culture, determine the character of organizational members, and helps to connect and bind people together in a cohesive team or family. Therefore, culture has the potential to place a considerable amount of influence on an organization because the shared values present within the culture are the important variables that determine and facilitate behaviors.

3.2.4 Symbols and Languages

Symbols are the things that are visible to members of the organizations that act as a reminder of the rules and beliefs of the culture. These reminders are important and necessary to help the members to maintain focus and be reminded of the rules implemented and supported by the management team. It also provides an insight into the way management will respond to some issues as they arise within the organization. The language spoken by the organization is a direct reflection on the way workers view their relationship with management and coworkers. Language can demonstrate trust, respect, and team cohesion.

3.2.5 Assumptions and Perceptions

The assumptions made by employees are usually based on information or the lack thereof that fuels the perception that they have of the management, coworkers, and the organization as a whole. Perceptions can impact employee's willingness to follow the leadership team and adapt to changes in business practices. The optimal position for management is when perceptions and reality are in alignment. When this occurs, the right stage is set and the correct message is heard and seen by everyone in the organization. Otherwise, the assumptions and perceptions of employees become the reality that mangers must deal with. Management should promptly clarify and address incorrect assumptions that employees may have to realign their perceptions of the workplace, the leadership team, and the organization.

3.2.6 Lean Thinking

Lean thinking is an integral part of a culture that is designed to support Lean process improvement initiatives and activities. Lean thinking represents a strategy accompanied by a way of thinking that is focused on the customer and continuous improvement in delivering quality products and services. In order for the goal of identifying and reducing waste in business practices to be realized, not only is it necessary for the leadership team to be Lean thinkers, there is a need for the entire workforce to be focused and embrace Lean thinking. Lean thinkers are focused on identifying and delivering value, creating optimal process flow, leveraging a pull system, and seeking perfection.

3.3 Subcultures

Often when we think of the culture of an organization, we look at the overall culture. What can be overlooked are the various subcultures operating within each group that form the makeup of the entire culture for the organization. These subcultures have the ability to impact groups in a way that can render them ineffective and significantly different from the rest of the organization. It is almost like they are a completely different city within a state.

These subcultures can take on different policies and practices and have their own language, values, assumptions, and perceptions. The workers in subcultures can even behave differently from other employees working in other parts of the organization. Before implementing Lean process improvement initiatives in a group or an organization, ensure that there is an understanding of the subculture of the groups being impacted. The subculture may indicate that process changes may not be embraced and therefore in jeopardy of achieving the desired level of success.

3.4 Establishing a Culture of Change

Many initiatives involving change ultimately fail, and the failure to sustain change once implemented is a reality over and over again. Significant change initiatives are successful when you have the right people onboard. The right people include those who are impacted and those who will be implementing the change. This can include workers as well as the leadership team. A flexible or culture of change is one that can embrace and facilitate change with some level of ease. This is not to say that there will not be some bumps in the road and there may be some level of resistance from workers. The bumps and the level of resistance are minimized because of cultural attributes that are prevalent and embedded that serves as a mechanism to get workers through the change process along with peer support from those who can readily adapt to the change.

The ability to successfully implement Lean requires that the culture of the organization be open to and able to adapt to change in conducting business. The lack of a culture that is supportive of change has contributed to the failure of many attempts to implement Lean process improvement initiatives. Establishing a culture that is supportive of change is the ultimate responsibility of the leadership team.

3.5 Change Management

It is incumbent upon managers beginning with the immediate supervisors to assist employees with adapting to change. For some, adapting to change is rather easy, and for others it is difficult to make the transition. To help with getting workers through the change process, management must be strategic and attentive to workers' questions, comments, actions, reactions, and needs.

Some actions that can be considered to facilitate change management are as follows:

- Know your culture and what it will tolerate.
- Ensure that the reason for the change is communicated.
- Identify workers that can help sell the change to others (*change agents*).
- Solicit buy-in and feedback by those who will be impacted by the change.
- Demonstrate full support for the change and communicate early with workers on the nature of the change and benefit to the organization.

3.6 Supervisor's Role in Change Management

The supervisor is the closest member of the leadership team to the employees; therefore, supervisors have an important role in the change management process. Supervisors should ensure that during their daily interactions with workers they are exhibiting the characteristics of a leader and a mentor. The supervisors can set the stage for how workers deal with change as well as help shape the culture so that the changes are easily embraced. To help facilitate change, supervisors should

- Be open and honest in communication.
- Provide a platform that employees feel free and safe to communicate their thoughts.
- Demonstrate support for the change.
- Never speak negatively about the change.
- Always highlight the benefits associated with the change while discussing how adverse or challenging issues will be addressed.
- Keep workers informed.
- Solicit feedback and input from workers.
- Be responsive to questions and issues as they arise.

Just as there are actions that are important for supervisors to engage in to facilitate change acceptance, there are actions that should be avoided that can stifle change acceptance. A short list of these actions is provided:

- Lack of supervision support for the change demonstrated verbally or through actions
- Failure to communicate openly and honestly
- Lack of follow-through with implementing the new technology, process, or procedure
- Failure to promptly answer questions and address issues

3.7 Assessing Culture

Before implementing Lean, it is a good idea to assess the culture of the organization to gage whether or not the culture is ready for and would be conducive for successful Lean implementation. This section will focus on tools that can be used to assess the culture of an organization or group.

Before beginning an assessment, ensure that employees are made aware of the pending assessment, the relevance and importance to the organization. Management should make a point to solicit open and honest feedback from employees when completing the survey. The following details should be provided to the workers before beginning the assessment:

- Why the assessment is being conducted and its importance in improving organizational performance?
- Time frame of when the assessment will take place.
- How the assessment will be performed?
- When the survey will be conducted?
- How and when the focus group discussions will be conducted?
- The process used to select participant for the interview portion of the assessment.
- What to expect during the focus group or individual interviews?
- How the result will be analyzed and used?
- When the results will be analyzed and when the results will be made available?
- Reassurance of anonymity and confidentiality.

It is also suggested that approximately 12 months after implementing Lean, conduct another culture assessment to gage cultural growth or degradation. A self-assessment of an organization's culture is an effective way to evaluate the attitudes, behavior, perceptions, practices, and policies that are in place that directly impact and guide the culture of an organization or group. There are three components that form the basis of culture, which are shown in Figure 3.3 in hierarchical order.

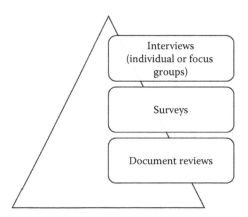

FIGURE 3.3
Culture assessment components.

3.8 Document Review

The survey or assessment should begin with a document review of written policies and procedures that govern the way work is performed and the expected performance and behavior of organization members. These documents should have clearly defined expectations to include roles, responsibilities, and accountability for members of the organization. If it is determined that expectations are ambiguous, this is the time to revise to ensure that expectations are clear to everyone across the board. Clear expectation is an important attribute that guides behaviors for members of the organization to include expectations for support for initiatives and functioning as an integral part of the team. Team work is an important organizational attribute when implementing Lean.

3.9 Conducting the Survey

Conducting a perception survey should be included as the second step of the assessment process. The knowledge gained from a perception survey can shed light on whether the organization or group is ready for the planned change or if management needs to focus on preparing the team for the change. Delaying implementation of Lean may be necessary during the preparation period. Preparation may include the following:

- Engage employees in developing the implementation strategy for the Lean process improvement initiative.
- Communicate the role that employees will play once the efficiency has been gained and Lean has been implemented. This action may help workers to remain through implementation and not *jump ship*.
- Reassure, if applicable, employees that no displacement of jobs will result from implementing the new process. If employees will be displaced, communicate what the process will entail and what support the company will provide in assisting them with future employment.

Conducting a perception survey will necessitate the selection of a survey instrument that has already been developed and validated or developing a new survey instrument. If a new survey instrument is developed, ensure that the survey is validated prior to usage. There are a host of survey instruments that have been developed to measure various attributes of culture. Please note that selection of the appropriate survey instrument is paramount.

Survey selection should be based on the type of feedback being sought and how the information will be used. Attributes to consider when selecting a survey instrument are as follows:

- Cost associated with procuring and distributing the survey.
- Does the survey have reliability and validity data? If so, is the data statistically sound?
- Does the survey ask the right questions?
- How will the survey be administered (electronic or paper)?
- Time required to complete the survey.
- Does the survey contain questions that are clearly understood?
- What method or methods will be used to analyze the data?

Survey development must also be based on the type of feedback being sought and how the information will be used. There are resources available to aid in survey development that may be used as a tool to ensure the survey is developed to complement its intended usage. Elements to consider when developing a survey instrument are as follows:

- What questions are needed to provide the level of information sought?
- Ensure that the survey questions are clear. Avoid ambiguous questions.
- Method used to demonstrate that the survey instrument is valid and reliable.
- What is the best method to use to administer the survey (paper, electronic)?
- What method will be used to analyze the data?
- Time required to complete the survey.

Table 3.1 contains examples of questions that may be included in a typical culture perception survey.

3.10 Focus Group and Individual Interviews

A focus group can consist of a group of people around 6–12 participants. The objective is to keep the group small enough to ensure that free flow of communication and information is facilitated. Focus group discussion is a valuable tool used to learn more about the opinions of organization

TABLE 3.1

Culture Survey Assessment Questions

1. Management demonstrates support for change.
2. The management team is trusted and respected by workers.
3. Employees enjoy their time at work.
4. Employees are motivated to perform their jobs.
5. Management is responsive and supports suggestions from workers.
6. In my organization, we celebrate our success and learn from our mistakes.
7. The management team is visible across the organization.
8. Communication from top managers is accurate when it reaches lower levels within the organization.
9. Morale is high across the organization.
10. My organization has a clear vision.
11. The vision of the organization is shared among organizational members.
12. My management is open and honest in communicating information.
13. Communication in my organization flows in all directions fluently.
14. The people in my organization easily adapt to changes in process and business practices.
15. I often recommend process improvement changes to my supervisor or manager.
16. Management is receptive to new ideas from employees.
17. Roles and responsibilities of organizational members are clearly defined, communicated, and understood.
18. My input is valued by my managers and peers.
19. My organization has clear goals and vision.
20. I believe that my opinion is valued by management.
21. Management has an open door policy.
22. Employees are motivated to perform their jobs.
23. The mission, values, and goals of the organization are posted for everyone to see.
24. Conflicts are handled fairly.
25. Knowledge and information are shared freely with everyone in the organization.
26. Disagreements are resolved quickly.

members on designated topics. Some guidelines to consider when conducting focus group discussions are as follows:

- Keep the list of questions short in the range of 10–15 questions.
- Ensure that questions are short and clear. Ensure the use of terms that are familiar to the participants.
- Use open-ended conversational type questions.
- Avoid using leading questions.
- Ensure that the questions are nonthreatening.
- Each question should focus on a single element or activity.
- Select participants who have knowledge of the topic.
- Select the appropriate meeting location that would ensure privacy for the group.

TABLE 3.2

Interview Question Examples

1. What do you enjoy most about working in your organization?
2. What words would you use to describe the culture of your organization?
3. What are the most common complaints workers have concerning your organization and leadership team?
4. How does communication flow in your organization?
5. Are you provided with the equipment and training needed to perform your job safely and successfully?
6. How comfortable are the people in your organization in communicating with management (supervisor, middle management, and senior manager)?
7. How comfortable are you in communicating with management (supervisor, middle management, and senior manager)?
8. How comfortable are you in raising concerns to management?
9. Are the skills and abilities of the people in my organization valued by management and used in their current job? How are they used?
10. How important is trust to the employees in your organization?
11. How are employees encouraged and willing to get involved in solving issues?
12. What words would you use to describe the management team?
13. How would you describe your supervisor's leadership abilities?
14. How does the leadership team demonstrate trustworthiness?
15. How often do you communicate your ideas and suggestions on improving business processes?
16. How do the people in your organization handle changes?
17. How engaged are the employees in your organization when it comes to participating in business outcomes?

Individual interviews can be used when gleaning information from one individual. This method is the best method to use to get feedback from senior managers. The same questions used during the focus group discussions should be discussed during the interview process along with additional questions that may be appropriate only for senior managers. Individual interviews should be kept to the prescheduled timeframe.

The use of focus groups and individual interviews is a good combination to collect additional data on the cultural health of an organization. Together the information gleaned can provide valuable insights into the views and opinions of the leadership team and the workforce. Examples of questions that can be explored during these interviews are listed in Table 3.2.

3.11 Shaping Culture

Many theorists and practitioners are of the belief that culture can be significantly changed by modifying various practices and the actions of the leadership team. Organizational culture comes together through combining various business practices that are shared by its members. There are several actions or practices that management can engage in to shape the culture of the organization to include the following:

- Change the way he or she communicates:
  - Open door policy
  - Communicate frequently and completely
  - Managers and employees share information openly, honestly, and frequently
- Change the things you value and the values of the organization.
- Management models the expected behavior.
- Create and implement a reward and recognition process that is fair and open to all.
- Demonstrate and promote team concept.
- Managers actively listen to workers.
- Managers demonstrate concern for employees well-being.
- Managers are visible and viewed as being helpful.
- Managers promote innovation and thinking outside of the box.
- Managers work together for a common goal.
- Managers create and share the vision for the organization.
- Managers consistently seek feedback from workers and act on the knowledge gained.
- Managers demonstrate trust in peers and subordinates.
- Managers support and embrace change.

Important actions to consider when attempting to shape culture are discussed in detail in Sections 3.11.1 through 3.11.8.

3.11.1 Communication between Workers and Management

Communication is simply the process used to transfer information through means such as verbal, nonverbal (body language), or written. The means by which we communicate can be one of the single most important attribute in the effort to shape culture. Workers expect management to communicate with them and keep them informed of business activities and protocol. When communication is not open, honest, and delivered with courtesy and respect, the information being delivered can be met with resistance. When implementing change, communication must be frequent and complete. Holding back on information that may leak out to employees by other means can negatively impact the success of change, the leadership team effectiveness, and the organization.

3.11.2 Fostering Trust from Within

Trust within organizations and between people has been studied by many practitioners and scholars. Collectively, they agree that trust is very important to organizational growth, productivity, and relationship building. The importance of trust in organizations was studied and recorded in the book titled *Culture and Trust in Technology-Driven Organizations*, Frances Alston, CRC Press, Taylor & Francis Group, 2014. The impact of trust in an organization and among its members cannot and should not be underestimated. People tend to follow those they trust even when things seem uncertain or unclear. There are several actions that management can take to foster trust in an organization. Some of these activities include the following:

- Always communicate frequently, openly, and honestly
- Treat everyone fairly
- Implement policies consistently across the organization as written
- Model the talk
- Demonstrate a concern for employee well-being and safety
- Be courteous and treat people with respect
- Demonstrate competence as a leader
- Freely admit mistakes when they are made

3.11.3 Consistency in Actions and Reactions

Employees depend on consistency in actions from their leadership team in addressing organization issues. This does not mean that the employees are expecting things to remain the same. Consistency can help enhance accountability, establishes the reputation of the leadership team, and add credibility to the message that is being delivered as well as the person delivering the message. A manager who is consistent in actions will be viewed as easier to trust and follow. The willingness of workers to follow management's directions and instructions is critical when implementing new technologies, processes, or policies. Followership is a critical factor in implementing and sustaining Lean process improvement and a critical factor in the change management process. Effective leaders are able to inspire people to follow.

3.11.4 Human Resource Policies and Practices

Human resource policies are a significant contributor that sets the stage for the culture of the organization. These policies can determine and often represent the mind-set of the management team through policy development and implementation. Once policies have been developed and approved, consistent implementation across the board is a must. In addition, management must himself or herself ensure that he or she is setting the expectations through being an example and adhering to the policies and procedures set by the organization. Workers in most cases follow the policies and procedures if they believe that the management team is the first partaker even if they do not fully support the policies and procedures.

3.11.5 Management Time in the Work Area

Many employees associate management time in the work area interacting with them as a sign that management is involved and is interested in their contribution to the success of the business. As such, management should spend time interacting with workers in their work area. While in the work area, management should take advantage of the time and discuss with the employees how work is being performed, evaluate whether the worker has access to the tools needed to perform his/her job efficiently and safely, and seek feedback from the worker on ways to improve the process, job task, and work environment. The feedback gained through these interactions can be valuable in the plight to continuously improve products and services.

3.11.6 Delegate Lower Level Decisions

For many managers, it is not easy to let go of responsibility and entrust them to someone else and trust that the task or activity will be performed to their satisfaction. As such, delegation is hard for some although not so hard for others. Delegation demonstrates a level of trust in another individual that can be instrumental in facilitating that individual extending trust in others. Also, delegation provides a level of empowerment among workers that can provide the springboard to confidence and the willingness to become engaged in business activities to include recommending and implementing Lean throughout the organization.

3.11.7 Employee Involvement

Companies that have a high level of employee involvement can expect implementing new programs and processes to be less challenging. Employees who are engaged in workplace activities are more committed to the goals and values of the organization. They are also motivated to contribute to the success of the organization. When it comes to adapting to change, these employees tend to adapt to changes easier than those who are not or less engaged. The importance and benefits of having employees who are engaged in the workplace and in Lean implementation are discussed in more detail in Chapter 6.

3.11.8 Leadership Charting the Way

The actions of leadership are the most important attribute in shaping culture. Leaders must own and lead the culture shaping process. Management can help shape culture through his or her actions and reactions. In doing so, there are some important actions and attributes that must be visible in leaders and the way they conduct business:

- Leaders must have a clear vision for the organization.
- Demonstrate the ability to be flexible and adapt to change.
- The value and expected behaviors of organizational members are modeled by the management team.

4

Employee Engagement in a Lean Culture

4.1 Introduction

More and more employees are spending a large proportion of their time at work with customers and colleagues. Employees who are working full time spend a large part of their lives in the workplace; how they feel about their workplace is of great importance to them and should therefore be of importance to management. The comradery of the workplace cannot be underestimated in its importance in driving behaviors and beliefs among workers. The connection that employees have among each other and with the organization can determine whether they are willing to become actively engaged in workplace activities or simply standing by and waiting for things to happen.

Many believe that active engagement can lead to the creative thinking that is necessary to improve processes and practices that can render a company more effective and profitable. Companies that have a high level of employee engagement often enjoy a culture that facilitates creative thinking and are able to embrace change and diversity. Also, in a culture where employees are actively engaged, there is a presence of a host of inputs or ideas on ways the company can improve in its business activities and processes. Without employee engagement, Lean implementation will not be effective or as effective, and it can and often has failed.

Some managers believe that it is their job to make all of the decisions, and an employee's job is to follow that unquestionably. This type of thinking is stifling to companies and their quest to be productive and create value for their customers. There have consistently been links between employees who are engaged in the workplace and the effectiveness of the support they provide to their companies to achieve their business objectives. These links have been demonstrated through research by various scholars and practitioners. Some of these links will be highlighted and referenced in this chapter. Employees who are engaged are instrumental in shaping the culture of an organization and can guide the success of the organization. Managers must keep in mind that in today's business environment, employee engagement can drive success or failure across the business.

4.2 Defining Employee Engagement

Many theorists and practitioners struggle with attaching a definition to employee engagement; however, many of them agree that having employees participate in workplace activities is of great importance for organizational success. When speaking of employee engagement, reference is being made to employees who are involved in providing inputs into decisions and having the ability to comment on the decision and suggestions of others. Based on the literature about employee engagement (Vignette 4.1), one can surmise that in order for employees to allow themselves to be engaged in work activities, there must be an emotional connection and an intellectual commitment with the organization that fuels the desire to help the organization succeed. In order to know how to address employee engagement in an organization, it is necessary to know what constitutes engagement by employees and how to encourage and improve engagement.

VIGNETTE 4.1

A literature review on the importance of various ways in which practitioners and theorists define employee engagement can be viewed in the following articles:

Markos, S. and Sandhya Sridevi, M., Employee engagement: The key to improving performance, *International Journal of Business and Management*, 5(12): 89–96, December 2010.
Sundary, B.K., Employee engagement: A driver of organizational effectiveness, *European Journal of Business and Management*, 3(8), 2011.

There are many books and articles that have attempted to define employee engagement in slightly different ways. Two such articles are referenced in Vignette 4.1 that provides a good resource on defining employee engagement along with some benefits of having employees involved in workplace activities. However, each definition suggests that input from and involvement of the worker is the premise behind what constitutes engagement. It is incumbent upon managers and supervisors to facilitate a work environment that is supportive of a culture that will encourage employees to connect with and allow themselves to become committed to the organization and its values and goals.

4.3 Emotional Connection in the Workplace

Emotions in the workplace play a significant role in the way the people in the organization communicate with each other, their external customers, and stakeholders. We have always heard that emotions and work should be kept separate. This is not realistic, because it is not easy to turn our emotions on and off. Emotions are a part of who we are and what we bring to the workplace. Emotional connection in the workplace is important for productivity. When companies create an environment where workers feel safe to explore and share new ideas, they will allow themselves to be concerned about the well-being of the organization. The concern developed among workers will facilitate engagement and will fuel the desire to connect employees with each other and with the goals of the organization.

A disengaged employee has a lack of emotional connection to the job, to his or her immediate supervisor or the leadership team, and to the company as a whole. Ways to encourage emotional connection in the work environment include the following:

- *Encourage relationships based on trust*: Emotional connection begins with trust.
- *A sense of belonging*: Feeling useful and valued by others is a great way to generate positive emotions and build relationships that will solidify employee's commitment to an organization's vision, goals, and values.
- *Realization of self-worth*: It is natural for employees that they want to know and understand their impact and contribution to the organization. A feeling of a sense of purpose can contribute to building a cohesive organization based on engagement and collaboration.
- *Encourage engagement*: Engaged workers are immersed in the business thereby allowing themselves to become emotionally committed. They have full involvement in the task at hand and are vested in the organization.

4.4 Employee Engagement Impacts and Benefits

A primary reason employee engagement is being discussed by all levels of management across the globe is the realized impacts and benefits. The benefits and impacts of having employees who are engaged in the workplace are yielding valuable benefits in the workplace. Many theorists and practitioners have surmised that employees who are engaged can and have directly added to the bottom-line of companies (Vignette 4.2).

VIGNETTE 4.2

Based on an extensive literature review, Solomon Markos and M. Sandhya concluded in their article entitled "Employee engagement: The key to improving performance":

The top drivers of employee engagement include communication between management and employee, providing growth opportunities for employees, and management being concerned about the well-being of workers.

Reference

Markos, S. and Sandhya Sridevi, M., Employee engagement: The key to improving performance, _International Journal of Business and Management_, 5(12): 53–59, December 2010.

There are five major benefits of employee engagement that are essential in getting work done and adapting to the changes associated with Lean implementation. These benefits are shown in Figure 4.1. Also shown in Figure 4.1 are characteristics and attributes that are important elements of each

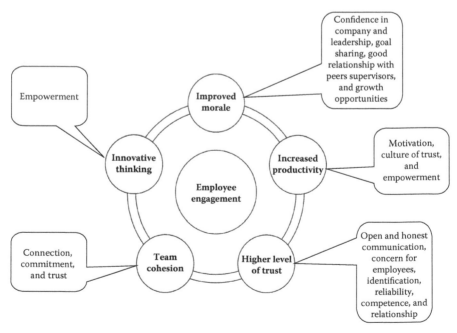

FIGURE 4.1
Employee engagement benefits and characteristics.

of the stated benefits. Engagement can impact an organization in many ways. These impacts have become a focus for many managers and practitioners.

Some additional benefits of employee engagement may include the following:

- Less absenteeism
- Fewer safety incidents
- Reduced employee turnover
- Increased creativity
- Commitment to the values of the organization

4.5 Improved Morale

Engaging employees in decision and policy making that impacts them directly is a great way to improve morale. Many times, managers will openly state that *our employees are our most important assets*; however, their actions demonstrate otherwise. Treating employees as if they are really an asset to the company is yet another way to increase morale and engagement. Improved morale has the potential to increase retention and the willingness to go the extra mile to ensure success of the company. The following are activities that have the potential to build and sustain high-employee morale:

- Be open and honest
- Be clear and transparent in communication
- Provide real-time feedback
- Celebrate accomplishments
- Communicate openly and frequently
- Build a culture of trust
- Show employees that they are appreciated

4.6 Employee Engagement and Increased Productivity

There is an increasing desire by executives and stakeholders for companies to increase productivity as a means to gain a competitive advantage in the marketplace and increase profits. It is not enough for a company to hire employees that show up for work every day and perform their task at the same pace using the same practices. There is an ever increasing expectation that productivity gains become a part of the continuous improvement process across the company. Employees who are engaged in the business can push the business to achieve more. Vignettes 4.1 and 4.2 provide key thoughts and beliefs on the impact of engaged employees on business performance.

Remaining productive is not just the responsibility of each employee, it is also the responsibility of the employer to create a supportive work environment that facilitates the desire to improve. When employees are involved and feel empowered to do their jobs, one can expect to see an increase in productivity. There are many ways to increase productivity in the workplace. The following list represents some ways that can increase productivity:

1. Avoid micromanaging
2. Empower workers to perform
3. Hold everyone accountable
4. Ensure that workers have the skills to perform their jobs
5. Set realistic goals
6. Recognize workers for performing good work
7. Follow up on activities promptly
8. Share knowledge and information
9. Avoid having meetings for meeting's sake
10. Build a cohesive team
11. Open and honest communication that is done frequently

Employees who are engaged in their jobs and work environment tend to be more productive because they generally are more motivated having the feeling of belonging and a sense of contributing to the success of the organization.

4.7 Team Cohesion

More and more teams are the predominant structure used to complete projects, activities, and tasks within the corporate environment. Team cohesion cannot become a reality without team members being actively engaged in supporting the team as a whole. A cohesive team is the first step in developing and sustaining a cohesive work environment that has many benefits such as increase employee involvement and satisfaction. Employees who are satisfied tend to participate and invest more in the work environment. In the workplace, it is pertinent that workers share a common goal to facilitate completion of tasks that is beneficial to the organization. Team cohesion begins with a group of individuals sharing the same goals such as completion of a project or a task. Sharing a common goal can promote inter-dependency among workers that can increase the ability to come together and form a bond that will increase team cohesion. Additionally, team cohesion can increase employee engagement and encourage employees to feel empowered to make decisions, and it also fosters better relationship between employees and with management.

4.8 Innovation and Engaged Employees

Innovative thoughts are sparked by employees who are involved in their work activities. When an employee believes that he or she is the key to the growth of the company, he or she tends to be willing to offer ideas and to engage in problem solving. Employees who are innovative thinkers seek novel and creative ways of completing tasks and work processes that can significantly improve business processes. Management should continuously encourage employees to share their ideas with their supervisor and coworkers. This sharing of ideas can spark dialogue of idea exchange that can lead to innovative thinking and innovative suggestions for process improvement. As a means to spark sharing of ideas, a suggestion box can be installed, and scheduling brainstorming sessions may also be helpful. It is clear that engaged employees can drive innovation in any work environment, and innovation can be accomplished through workers at every level when they are engaged in the business. It is a mistake to think that innovation is confined to a research and development environment or by workers who are highly educated and experienced.

4.9 Higher Level of Trust Contributes to Engagement

Trust is unquestionably one of the single most important elements that can improve employee engagement. When people trust, they are willing to allow themselves to be open to different thoughts and ideas and are willing to follow the direction of others. Trust also provides the thrust that is needed by individuals to allow themselves to be influenced by another person. A person with influence has the power to guide and direct the actions of others. Influence can be a powerful tool in organizations that are implementing new technologies or embarking upon a change in process or procedure. There are many papers and books written on the importance of trust in organization. Two suggested readings are listed in Vignette 4.3.

VIGNETTE 4.3

Shockley-Zalabak, S. and Morreale, S., *Building the High-Trust Organization: Strategies for Supporting Five Key Dimensions of Trust*, J-B International Association of Business Communicators, San Francisco, CA, 2010.

Alston, F., *Culture and Trust in Technology-Driven Organizations*, CRC Press, Boca Raton, FL, 2014.

Trust in leadership and the organization is necessary when implementing new processes such as Lean. Trust opens the door for creative thinking, team cohesiveness, leadership acceptance, and the desire to follow.

4.10 The Actions of an Engaged Leader

The philosophy and the actions of the leadership team are integral in developing and maintaining a culture where employees are actively involved in activities that will add to the success of the company. Not only is it necessary for employees to be engaged, they must feel a connection with the organization and the leadership team. This feeling of being connected will motivate them to buy-in to changes easier and help the management team in facilitating changes. When implementing Lean, it is important to have the employees as well as the leader engaged. The engaged leader is the key and helpful in setting the stage for a culture that encourages and facilitates employee interaction. Some characteristics of a manager who facilitates an environment that encourages engagement include the following:

- Seeks worker feedback
- An active and effective listener
- Provides prompt feedback on questions and issues raised
- Communicates openly
- Leads with humility

Leaders who understand the importance of engagement among workers will take the time to become a student of the philosophy. Engaged leaders inspire employees to become actively engaged employees. These leaders take the necessary time needed to educate themselves on ways to cultivate an environment that will enhance the feasibility of employees becoming interested enough in their workplace to become an active participant.

4.11 Communication Strategy

Communication skills are critical in every interaction at all levels on a daily basis, and communicating effectively in the workplace is critical to the success of business whether communicating with coworkers, colleagues, or customers. Individuals who are informed are generally more trusting of their management, colleagues, and confident in their contribution to the business. On the other hand, in an environment where poor communication exists, team members can become disgruntled, deadline can be missed, and there can be a potential increase in employee turnover. When communication flows, knowledge is shared and issues are addressed. When a company does not communicate well, information becomes stove piped and does not flow freely. In this type of environment, the rumor mill is quite active and potentially disruptive.

Each organization should have a communication strategy that describes the rules of engagement when it comes to various types of communication within the company as well as external to the company. It is just as important to have a strategy to communicate how Lean will be implemented within the organization and the need for employees to be actively engaged in the process.

4.12 Human Resource Policies and Practices

A critical important issue in Lean success in gaining the focus of leaders is the relationship between the human resources (HR) policies and practices and Lean implementation. These policies and procedures can serve as a tool that supports the process or a barrier to process improvement. Human resource policies set the stage for whether an employee believes that engagement in the organization is supported and safe. These policies can provide empowerment to employees when it comes to giving them a feeling of safety in giving open and honest feedback to management on innovative ways to improve process and on issues as they arise. The HR department interacts virtually with every part of the company, which makes it an ideal group to be a powerful ally in Lean implementation. For example, The HR department can help in conducting training sessions and awareness campaigns for employees. It is very important that employees understand how to perform their roles after process changes have taken place. One of the most important things that must happen in implementing Lean is the awareness of workers on the activities taking place in the organization. It is essential that the HR department actively supports all aspects of Lean transformation. A list of the critical roles that the HR department may perform that can impact Lean implementation is documented in Table 4.1.

TABLE 4.1

Human Resources Role in Lean

Activity	Purpose	Comment
Training	Train employees on the new process or technology or training to be able to take on another role in the company	There are times when implementing Lean, the end result is that employees may be displaced because systems and processes are operating more efficiently and therefore require less manpower to operate
Job placement	Provide different assignment for displaced employees	Employees who are displaced by Lean implementation will be in need of securing other positions. It is preferable if the position is internal to the company. Internal job placement put at ease other employees when it comes to Lean implementation
Communication	Keep current employees on changes and how and if they will be impacted	Communication can relieve the fear of the unknown. It is normal for people to be concerned with how implementing a new efficient process will impact them and potentially their family
Job counseling	Provide assurance to employees that they will have a role in the company if feasible. In the case where an employee will be displaced, provide him or her information on what the company is willing to do to assist him or her in locating employment internal or external to the company	Displacement of employees at times is a reality when implementing Lean, if not addressed adequately, a lack of trust and fear of uncertainty can result
Policy and procedure development	The new process may result in a change to policy or procedures	Ensure that policies and procedures are modified early in the process and employees are briefed

4.13 Evaluating Employee Engagement

Before implementing Lean, it may be valuable to evaluate the level of employee engagement found within the organization. The knowledge gained from this evaluation can be instrumental in assisting with the strategy to implement Lean. In addition, the knowledge gained can serve as a tool to aid in modification of the Lean implementation strategy and in the way feedback is solicited and provided to and received from employees.

There are two primary means by which employees can be surveyed to determine if the appropriate level of employee engagement is available to successfully implement a new process, procedure, or technology. The two means are through surveys or focus group discussions. We will discuss the two methods in more detail in Sections 4.13.1 and 4.13.2.

4.13.1 Employee Engagement Surveys

Today employers use employee engagement surveys to measure employee beliefs or perceptions on things such as passion for work, the value and inclusion of their ideas, and ideals by management as it pertains to getting the work done among other elements. Many companies frequently administer employee engagement surveys on a routine basis such as annually. As a result, there are many such surveys available for use. Some of these surveys have been validated and have reliability data to demonstrate survey repeatability. Care must be taken when choosing a survey that has been developed versus developing your own survey. An important aspect of survey selection or development is to ensure that the appropriate questions are asked. A set of questions that can be considered for use in an employee engagement survey is included in Table 4.2, and a survey questionnaire is documented in Chapter 11.

4.13.2 Focus Group Discussions: Employee Engagement

Conducting focus group discussions is an essential compliment to administering a survey or a questionnaire. The information gained through these discussions can provide insights into the following:

- Are employees engaged in the work environment?
- Do employees feel safe in allowing themselves to be engaged?
- Does the work environment facilitate engagement?
- Does the management team support engagement of employees?

A focus group generally is most effective if it consists of approximately no more than 6–10 people brought together to engage in a discussion guided by a facilitator to provide information on specific topics. Before holding the focus group session, ensure that the objective of the focus group has been developed, the time frame for holding the session has been determined to include dates and time, the questions to be explored have been developed, participants have been identified, and a facilitator has been appointed.

TABLE 4.2

Employee Engagement Survey Questions—Example

---

1. My management listens attentively to my suggestions.
2. I am willing to participate in a team to implement a new program or process.
3. My manager values my opinion.
4. My manager often seeks out my opinion.
5. I am proud to tell people where I work.
6. I have the opportunity to participate in decisions that affect me and the organization.
7. I am willing to provide feedback to my management on issues.
8. I feel valued by my managers and teammates.
9. My management inspires me to be engaged in organizational activities.
10. I enjoy coming to work.
11. I feel valued for the work I perform.
12. My supervisor provides guidance when I need it.
13. Management provides the tool I need to do my job.
14. I know how the work I perform fits into the strategy of the organization.
15. Employees are motivated to perform their work.
16. Management holds himself or herself and everyone in the organization accountable.
17. Employees in my organization take responsibility for their decisions.
18. My workgroup works together as well as a team.
19. I am supportive of changes that help me do my job more efficiently.
20. I enjoy working with my team.
21. I have the skills I need to perform my job effectively.
22. I speak highly of my supervisor.
23. I am satisfied with my job.
24. I am committed to doing a quality job.
25. I am constantly seeking ways to do my job better.
26. My supervisor expects me to work safely and look out for my coworkers' safety.
27. I would recommend my organization to others as a great place to work.
28. I am willing to put in extra effort to get my job done when needed.
29. My management is concerned with my safety.
30. I willingly offer my assistance to others who have heavy work load.
31. I am concerned with the safety of myself and my coworkers.
32. My talents are utilized by my organization.
33. I trust the accuracy of the information I receive.
34. I know what my supervisor expects of me while at work.
35. The days I look forward to going to work outnumber the days I do not look forward to going to work.

When developing the questions for a focus group, ensure that the questions are designed to yield the desired results. Use well-designed and open-ended questions and avoid using yes or no type questions. Examples of good focus group questions are shown in Table 4.3.

TABLE 4.3

Employee Engagement Focus Group Questions—Example

1. What five words would you use to describe your leadership team?
2. How does your supervisor encourage you to be engaged in helping the organization to solve problems?
3. How important is trust to the members of your workgroup?
4. How does management communicate with employees in your organization?
5. What do you like the most about working in your organization?
6. How are organizational changes communicated and implemented by the leadership team?
7. What type of equipment and training has been provided to you to perform your job successfully?
8. How engaged are you in making decisions that impact you and your coworkers?
9. What actions have your supervisors performed to inspire you?
10. How engaged is the management team in ensuring that the voice of the worker is heard?
11. How does management communicate changes in process, technology, or procedures?
12. How receptive are employees when management communicates changes in process, technology, or procedure?
13. How do you feel about coming to work each day?
14. How are employees encouraged and willing to get involved in solving problems?
15. Describe the flow of communication in your organization.
16. What are the most common complaints that workers have expressed concerning the organization and the leadership team?
17. How comfortable are you in communicating issues to management (supervisor, middle management, and senior management)?

4.14 Summary

Employee engagement and involvement is a management philosophy that engages employees' contribution in continuously improving business practices and processes. One thing for sure is that management of all types competing in various industries agrees that employee engagement can provide a competitive advantage for a company. This philosophy is embraced because of past experience of management and theorists and in their study and observation of the benefits of engagement. If one were to research studies on engagement and productivity, it would be clear that engaged employees are often more productive. Not only are they more productive, they will most likely remain with the company for a longer period of time, they will allow themselves to be vulnerable enough to trust management and the organization, morale is usually higher, and they are more creative and willing to offer ideas to improve business processes and practices. Positive engagement of this type contributes immensely to business outcomes and the bottom-line.

5

Succession Planning Strategy in a Lean Environment

5.1 Introduction

Succession planning is an important part of a corporate business strategy that can often be challenging to implement in an efficient manner. Succession planning recognizes the talent and skill set needs of a company to ensure that business processes can continue operations into the future. It also recognizes that the top talent people of the present can be the top talent people of the future working in positions of more authority. It is a viable process that is used to help companies manage the future of their top talent. Growing talent from within is a significant step in building bench strength for companies. The inability to implement a good strategy that can withstand the test of time can result from the following:

- Inability to maintain the skilled staffing needed.
- Strategy is too aggressive to implement.
- The strategy is written on paper and not communicated.
- Lack of leadership support and commitment.
- Lack of buy-in from employees.

Another activity that can be added to this list is the implementation of Lean thinking and process improvement tools. Once Lean process improvement enters the picture, employees often will begin to ask questions such as the following:

- Is the company in trouble and in need of a reducing cost?
- Are we heading for a layoff?
- Are our jobs stable?
- Will my position be retooled to the point that I will no longer be valuable to the company and the process?
- Will I be laid off?
- Without a job, what will happen to my family?

The benefits of implementing Lean within a company is highly recognized and embraced by corporate leaders. Numerous companies have cited benefits at all levels. However, many times Lean improvement initiatives are not always successful because the people aspects of the process are not addressed early on in the process. Succession planning is a key business process that can be significantly impacted by implementing Lean if not focused upon.

There are various changes in business processes that will facilitate the need for succession planning. These activities can include the following:

- Changes in corporate business strategy
- Changes in technology
- Knowledge retention
- High-employee turnover
- Implementation of process changes
- Restructuring of the organization

Managers in various disciplines and fields depend on succession planning to ensure the availability of technically competent workers. Engineering managers also depend heavily on succession planning activities to ensure the availability of highly skilled engineers to support projects that are necessary to fulfill their customer requirements. No matter what type of business you are engaged in or leading, succession planning is an integral part in achieving goals and protecting the bottom line for the future.

5.2 What Is Succession Planning?

Succession planning is an important process that is used to increase the availability of primarily internal candidates to fill key positions as they become available within a company. The most simplistic definition of succession planning is shown in Figure 5.1.

A succession planning is an integral part of a company's overall business strategy and should be included in the human resource process. Often succession planning is used to target only senior executive positions. However, all key positions at all levels of the company should be included in the plan. Effective succession planning does not begin and end with identifying the candidate as is often seen. The process is only complete when the candidate has been developed and provided with the appropriate tools for success and is able to function in the identified position. Vignette 5.1 provides an example of what can happen when key positions are not included

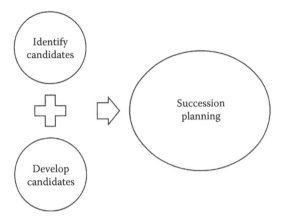

FIGURE 5.1
Succession planning components.

in the company's succession planning or when there is no strategy to fill critical positions within a company.

VIGNETTE 5.1

Company A is a global company with several key executives who are instrumental in supporting operations and sales. The executive responsible for Asia took a position with their competitor. The CEO felt that it was not necessary to include this position in the succession planning strategy because of the added cost associated with training and preparing a candidate. He further stated that the training budget was significantly reduced, and succession planning activities were greatly reduced to cover only critical senior level positions since the implementation of Lean process improvement initiatives. Having no one familiar with the Asia business and customer base and having no one to fill the position on an interim basis, customer care and feeding is on hold until a new manager is in place.

Succession planning is used as a tool by companies for one or more of the following reasons:

- To develop people and strengthen their ability to advance
- To have the right leaders prepared for the right position at the right time
- To keep key positions filled to ensure that the overall corporate strategy is implemented
- To preserve the performance of the company
- To meet the expectations of investors and stakeholders

5.3 Developing a Succession Planning Strategy

A comprehensive succession planning strategy is an important component of the overall business strategy for a corporation to meet its goals. An effective plan has the right mixture and the appropriate level of management support, the right candidates to fill key positions, and a strategy and plan to develop each successor (Figure 5.2). If there is an unbalance in this combination of elements, the strategy has the potential to be unsuccessful in meeting the needs of the organization.

Succession planning is an ongoing process that will require adjustments as the organization changes and implements new product lines or processes. Often succession planning is placed as a responsibility of the human resources department. The strategy should be linked to the company's human resource goals and objectives but should not be made a primary responsibility of the human resource department. The responsibility for succession planning must reside with the entire management team. A succession strategy will not be successful without management involvement and commitment.

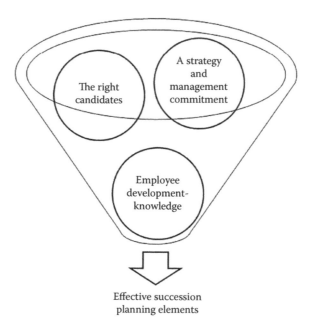

Effective succession
planning elements

FIGURE 5.2
Succession planning elements.

5.4 What Does a Good Succession Strategy Look Like?

A good succession plan is developed with the current and future needs of the business in mind. With the explicit purpose of seeking out and preparing potential successors, the plan should include the following elements at a minimum:

- A list of positions
- A written procedure that provides the criteria for identifying the right candidate
- Competencies required for key positions (both technical and leadership skills)
- A list of positions that are hard to fill
- A list of positions that may require internal posting
- A list of successor candidates
- A method used to conduct the gap analysis
- A process used to close or fill gaps in knowledge and skills of the successor candidate

A pictorial view of the succession planning process is shown in Figure 5.3. The diagram is based on the model developed and discussed in *Guide to Environment Safety and Health Management* (Alston, 2015). The model has been revised to include an important element that is not specifically called out in that model. The additional attributes that must be focused on separately are positions that are hard to fill. Based on experience, many managers have come to realize that there are often positions that are not easy to fill internally for various reasons. Special considerations must be given to these positions with the knowledge that it may include adding to the strategy a provision for including an external candidate search.

Some reasons why some positions may not be easy to plan for through internal selection include the following:

- Inadequate skill mix
- Limited employees with leadership potential that can be developed in a timely manner to fill a key position
- Lack of interest by employees for the key position

Some reasons why some positions may not be easy to plan for through external selection include the following:

- The existing period is of a good economic growth.
- Company benefit package is not competitive.

- Company is located in a remote area.
- Company is located in a high cost of living area.

To further elaborate on the succession planning model, each of the steps is further defined in Figure 5.3 in detail beginning with the first step in the process.

5.4.1 Identify Key Positions

The first step in developing a succession strategy is to identify the positions that are the key to the success of the organization. This would require an extensive analysis of the company's current business strategy as well as the goals of the organization for at least the next five years. Selection of key positions may be assigned to an individual or a committee to identify the positions and to select candidates. During the selection process, care must be taken to prevent office politics that may interfere with selecting the right successor candidate. Key positions can be found at all levels in the company in essentially every position. Several avenues

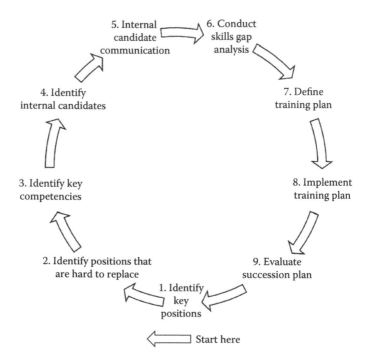

FIGURE 5.3
Succession strategy model.

can be used to select the right candidate for each position. Some of those avenues are as follows:

- Board selection method through a documented nomination process
- Individual nomination with board selection based on documented selection criteria reviewed by human resource and possibly by the legal department
- Internal board interview process
- Recommendation by management

A table such as Table 5.1 serves as a starting point that can be used in collecting your thoughts and cataloging them.

In order to ensure that all key positions are appropriately identified, management at all levels must be involved in populating the list such as the one in Table 5.1. It is not always obvious what position is the key when viewed by someone outside of the organization. Therefore, the manager for each group or department must be actively involved in the process.

5.4.2 Identify Positions That Are Hard to Fill

At times, there may be key positions that are difficult to fill and may need extra attention or extensive strategy and coordination to fill when they become vacant. These positions can present some challenges, and there may not be any viable candidates for internal targeting. It may be necessary to fill

TABLE 5.1

Key Position Catalog

Position	Department	Responsible Manager or Committee
President		
CEO		
Supervisor		
Vice president (VP)		
Human resource manager		
Health physicists		
Sales		
Communicator		
VP environment safety and health		
Research analysis		
Executive assistant		
Quality engineer		
Engineering manager		

these positions through external means. In such cases, it is imperative that these positions should be included in the strategy and communicated why it may be necessary to fill externally some, most, or all of the time. It is important to understand why these positions are identified, what makes them difficult to fill, and how they will be filled. When considering external candidates, it is critical that each candidate is evaluated against a set of defined criteria that can provide screening such that the right candidate is selected. The succession strategy should include an evaluation of those positions that are slated for external hire and why they are external hire positions. The progression of such a decision should be documented and communicated.

There are various actions that can be taken and various resources that can be instrumental in implementing succession planning for these positions. The most commonly used avenues are listed in Figure 5.4.

5.4.3 Identify Key Competencies

Competencies are those skills, knowledge, and attributes needed by a candidate to be successful in a particular position. When identifying key competencies, ensure that all of the skills and knowledge that are required to successfully function and add value are identified. Be careful in disregarding any competency without careful, even if the selected successor does not possess a required competency. An example of a worksheet that can be used to capture thoughts and the process is shown in Table 5.2. Be careful not to develop key competencies based on the attributes of a preselected candidate. This action will render the identification of competencies invalid or weak in achieving the desired outcome.

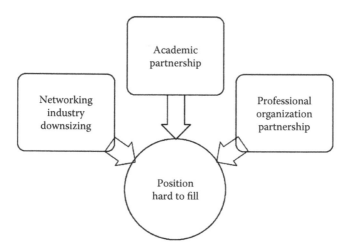

FIGURE 5.4
External sources.

TABLE 5.2

Key Competencies Worksheet—Example

Position	Critical Skills
President	• Leadership: Demonstrated skills in leading a staff of 100 • Budget management: Management of a budget of more than one billion
CEO	
Supervisor	
Vice president	
Human resource manager	
Health physicists	
Sales	
Communicator	
VP environment safety and health	
Research analysis	
Executive assistant	

5.4.4 Identify Candidates

The process of identifying candidates for a current and future critical position begins by looking across the company and by identifying the potential candidate pool. Some plans may include only *high potential* or *top talent* employees. Other plans are developed so that anyone in the company can participate and are included in the candidate pool if they meet the defined criteria. Many times, managers seek out people who are just like them. For some managers, this seems to be an automatic process when selecting individuals for promotion and leadership-type positions. This action should be avoided and the manager should stick to identifying candidates based only on the defined criteria and position description. A diverse candidate pool provides the basis for an infusion of new ideas and ideals into the organization that can improve performance across the board. When selecting succession candidates, care must be taken to ensure that the process is viewed as being fair by employees. The selection process should be objective, critical, fair, and seamless. A plan that is not viewed by workers as affording opportunities to all who are qualified can negatively impact employee morale, trust in the leadership team, and the overall culture of the organization. Behaviors to avoid when selecting candidates include the following:

- Avoid force fitting a square peg into a round hole. For example, avoid selecting an engineer to take on a role as an engineering manager just because they are viewed as great engineers if they do not have the ability to lead.
- Do not select candidates solely based on the opinion that they have been loyal to the company. A loyal employee does not necessarily make a good leader or manager.

During the selection process, it is possible to identify more than one candidate for any given position. Selecting more than one candidate for a position is desirable; however, this is not always feasible especially in organizations where Lean process improvements have been initiated. Often in Lean organizations, the amount of resources is kept at a minimum, which makes it challenging to identify and prepare more than one candidate for a prospective position (Table 5.3).

5.4.5 Internal Candidate Communication

Once the internal candidates have been selected, management should communicate to the candidate that they have been selected to participate in the company's succession planning process. The candidate should be made aware of the selection process used and the role the company is seeking for him or her to fill in the future. If feasible, provide the candidate a time frame on when he or she should be expected to assume the new role. Also, this is the time to get feedback from the candidate on acceptance or questions he or she may have on the selection process or the position in which he or she was named as a successor. In addition, use this time to enquire and validate the employee's experience and gather information that will be used to develop the gap analysis and training plan.

5.4.6 Conduct Skill Gap Analysis

A skill gap analysis is designed to capture the areas for growth required to prepare the successor for the position he or she is expected to fill in the

TABLE 5.3

Candidate Pool Identification Worksheet—Example

Position	Date Position Available	Candidate 1 Name	Candidate 2 Name	Candidate 3 Name
President				
CEO				
Supervisor				
Vice president				
Human resource manager				
Health physicists				
Sales				
Communicator				
VP environment safety and health				
Research analyst				
Executive assistant				

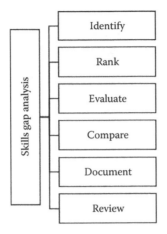

FIGURE 5.5
Gap analysis process.

future. This step is the first step in preparing the successor for success. A comprehensive gap analysis focuses not only on the technical skills but also on the soft and leadership skills complementary to the position. Basic elements of developing a gap analysis to determine delta in skills are shown in Figure 5.5 and are described as follows:

1. Identify the skills needed to achieve success in the position in its entirety
2. Rank the required skills in order of importance
3. Evaluate the skills of the candidate
4. Compare the skills of the candidate against the skills needed for the position
5. Document the gap in knowledge and skills
6. Review the gap analysis with the candidate

The information gathered from conducting the gap analysis can be summarized in a skill analysis work sheet. An example of a typical work sheet is shown in Table 5.4.

5.4.7 Define and Implement Training Plan

A very comprehensive and implementable training plan is an essential aspect of the succession planning strategy. Developing a plan to fill the gaps in knowledge and skills for the successors can be accomplished through various means. However, management involvement and commitment from both management and the candidate are essential. The training plan is a personalized plan to address specific gaps for each candidate. The plan is developed from the knowledge gained from the skill gap analysis. There are

TABLE 5.4

Candidate Skill Analysis Work Sheet—Example

Position	Critical Skills	C1	C2	C3	Delta in Skills
President	Leadership, financial, and strategic planning	Sally Doe	Jane Doe	John Doe	C1—Leadership training C2—Communication C3—Experience
CEO					
Engineering manager					
Vice president					
Human resource manager					
Health physicists					
Sales					
Communicator					
VP environment safety and health					
Research analyst					
Executive assistant					
Quality engineer					

C, Candidate.

various paths that can be taken to close the gap identified in the gap analysis. The training plan should contain at a minimum the following:

- The skills that must be acquired or sharpened
- The process used to acquire the necessary skills such as formal classroom training, mentoring, and coaching
- The projected time frame necessary to close gaps

The plan and progress should be reviewed at some frequency with the candidate especially if completing the plan is scheduled for a period of more than six months to a year. The plan can be modified if additional gaps are discovered during the training phase.

5.5 Promoting from Within

Promoting from within is a management tool to leverage internal talent for critical position within the company. It states to employees that the company is willing to invest in them and that they are a valued resource.

Some of the reasons why companies promote from within are as follows:

- Promoting from within is generally less costly than external hires. The company can save on advertisement cost, moving cost, and the associated cost of onboarding a new employee.
- An inside hire is already acquainted with the culture, policies, and procedures of the company.
- An inside hire already has a relationship with internal employees and should be able to become productive in the new position sooner.
- Promoting from within serves as a motivator for other members of the staff to work harder in order to be promoted.
- Promoting from within can produce multiple vacancies because the individual promoted would most likely be replaced.
- Promoting from within can serve as a retention tool for employees.
- Promoting from within can increase employee loyalty.

When promoting from within, keep in mind that just because an employee is skilled and experienced in his or her current job does not necessarily mean that he or she will be successful as a manager or leader. A skill assessment should be performed to assess the soft and leadership skills needed to ensure that the selected candidate will have what it takes to be successful.

Some of complexities that one may encounter when designing and implementing a strategy in support of promoting from within include the following:

- There must be a training strategy to prepare candidates.
- May facilitate hiring of unqualified candidates if based on politics.
- Can deplete morale of workers if there is a perception that unqualified candidates are being promoted.
- May limit the infusion of new ideas.
- Difficulty to change culture.

To sum it up, a company that has implemented a policy of hiring from within can enhance the feasibility of employees challenging themselves at all levels because they believe that there is an avenue for career growth within the company. This can result in increased loyalty and reduced turnover among employees. Table 5.5 summarizes the potential advantages and disadvantages of hiring from within versus hiring externally.

TABLE 5.5

Internal versus External Hiring

Promoting from within	External Recruiting and Hiring
Advantages	
• Improved employee morale. • Improved loyalty to the company. • Lest costly. • Easier to integrate into the culture of the organization. • Easier to assess applicant's true ability since the candidate is a known entity. • Promoted employees are already familiar with the organization culture, policies, and mission. • Demonstrate to other employees that career growth is feasible within the company.	• Provides an infusion of new ideas. • Can reduce the amount of training needed since the employee would have experience. • Increase diversity. • Add more talent to the talent pool.
Disadvantages	
• Training costs will incur. • Learning curve may be steep. • Smaller talent pool. • Internal politics may interfere with hiring the most qualified candidate. • With a smaller talent pool, it may be difficult to comply with internal policies such as affirmative action. • Will have to replace vacated position.	• Less knowledge of the capability of the candidate. • Candidate search process is longer and more costly. • Resistance of new ideas by current organization members. • Employees may lose faith in their ability to advance.

5.6 Knowledge Gathering and Retention

Knowledge retention is another good reason to have an effective succession planning strategy. Knowledge gathering and retention involve capturing knowledge possessed by organizational members so that it can be used by others later when needed. Knowledge must be captured periodically and transferred to other successors periodically for success of the organization. Knowledge retention must be a critical part of the succession strategy. It can be critical to the success of a project, product line, or a business. Yet many companies do not have a knowledge retention strategy and many times loose access to critical information when workers relinquish their ties with the company. Knowledge retention is discussed in further detail in Chapter 6.

Vignette 5.2 serves as a reminder of what can happen if a knowledge management plan is not implemented within an organization.

VIGNETTE 5.2

John served as a product distribution manager for his company for more than 10 years. The company implemented Lean process improvement, and John's position was consolidated with another position. John was informed that he was no longer needed. During John's exit interview, his manager and the human relations representative tried to use that time to gather knowledge from John concerning the many unique clients he served. At the end of the interview session, John walks away unsure of whether he had remembered to disclose all of the needed critical information.

It has been established that employee turnover can cost a company in many ways to include the cost associated with onboarding new employees and low morale for the employees remaining with the company.

When developing the knowledge retention strategy, there are three questions that must be posed before developing the plan. These thought provoking questions are listed in Figure 5.6.

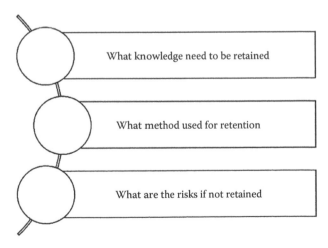

FIGURE 5.6
Knowledge retention questions.

5.7 Continuing Education

Once the candidate has completed the training plan, he or she should be included in the company's continuing education process. It is necessary to help maintain and build on the candidate skills, provide a means of networking, and learn new concepts. Continuing education can be accomplished through one or more of the following:

- Classroom training
- Participation on boards or committees
- Conference attendance
- Reading books
- Participation in workshops

5.8 Succession Planning in a Lean Environment

Succession planning is recognized as a valuable process used by forward thinking companies to take care of the company into the future. It is also recognized that the process is not easy to implement especially when implemented in a Lean environment. I am defining a Lean environment as a process or organization where Lean has been implemented to improve systems and streamline complexities and redundancy in efforts. Generally, in these environments, the level of staffing decreases once efficiencies are realized. Succession planning is easiest when there is a certain level of available bench strength. Lean process and tools are generally used to eliminate the need for exhaustive processes and practices, and thus can eliminate some level of bench strength that can make succession planning easier. Therefore, managers must be able to pull upon their creativity and think outside of the box to ensure that effective succession planning is accomplished. Some activities that managers should consider to ensure that succession planning is effective when Lean thinking and process improvement is a part of the organization's culture can include the following:

1. Have the candidate work in an alternate work shift if feasible. Such as a 9/80 shift which will afford them every other Friday away from the work environment. On the scheduled days away from work, the employee would work on completing his or her training strategy to prepare for the position in which he or she has been slated for. This method can add an additional cost because the employee may be paid extra for the training time, and it can also add to employee

stress and fatigue because he or she will be working additional hours within the workweek.

2. Although it may be difficult, carve out some time during the workweek for the employee to train and prepare. Recognizing that this may have an impact on production and schedules.

3. Reassign the employee to a task or position that will afford some downtime during the day to complete the training plan.

4. Relieve the employee from the day-to-day work scope and allow him or her to work full time on the training plan. This method may require the need for additional resources to cover the work scope that the employee is not completing while on full time training, or other employees to take on more work and work longer hours each day. There will be a cost associated with this method and the potential of addition stress on employees. The following scenario listed in Vignette 5.3 is a good example of possible outcomes when implementing Lean and potential impacts on succession planning.

VIGNETTE 5.3

Jim is always encouraging his managers to *think Lean* when it comes to process improvement. One of his managers implemented Lean process improvement for the mass production assembly line that produced specialized parts for the automobile industry. As a result of the initiative, he was able to improve throughput to the point that the amount of resources needed to operate a specialized piece of equipment went from three workers per batch to two workers per batch. To avoid laying off a good worker, one worker was reassigned to another process that was short staffed. The workers were not involved in the development or implementation of the new Lean strategy. The two remaining workers are now questioning the stability of their jobs and are actively seeking work outside of the company.

Thought provoking question: How did management actions impact succession planning?

When implementing Lean, consideration should be given to the impact that it will, and can, have on succession planning. Just because implementing Lean may add additional complexities to succession planning does not mean that Lean is not the right path to take. Just remember to revisit your succession strategy when implementing Lean early on in the process. Early reviews allow the management team to begin considering knowledge retention and the needs of the company when it comes to retaining skilled workers and leaders to maintain the viability and productivity of the company.

5.9 Summary

Succession planning is a must for companies to continue to achieve a competitive edge in the business sector in which they compete. Planning for the future gets the organization out of the reactive mode and places them into a proactive mode of strategizing for the future of the company. Recognizing that with the implementation of Lean thinking and tools, planning for the company's future when it comes to having the right staff with the right skill set may be a challenge that must be overcome. Overcoming these challenges will require management to be strategic in planning and implementing Lean. The people aspect of Lean implementation must be considered early on in the process.

Succession planning is not a one-time event rather an ongoing activity; the plan should be evaluated at some periodicity and when there are changes within the business structure. Changes such as influx of personnel, high staff turnover, or the company business strategy are revised. A good rule of thumb would be to evaluate the plan annually. However, in a more stable business environment, evaluation of the plan every other year may be acceptable. A good succession strategy can assist in reducing turnover and in shaping the overall cultural environment of a company.

Reference

Alston, F. and Millikin, E.J. Guide to environment safety & health management, developing, implementing, and maintaining a continuous improvement program. Boca Raton, FL: CRC Press, Taylor & Francis Group, 2016.

6

Talent Management and Retention and the Hidden Costs

6.1 Introduction

Developing and maintaining an environment where employees are satisfied and are actively participating in the business take planning and constant attention to changes in the work environment. All changes in the work environment have the potential to impact workers in various ways. Some changes may be welcoming and others may not be. These changes can serve to fuel an employee's thought of looking for another place of employment. Talent retention can be a daunting problem for companies and can impact productivity, retention of key resources, and the financial bottom line. Replacing existing employees cannot always be quantified in dollars and cents because it involves two different types of costs often referred to as direct and indirect. Direct costs are easiest to calculate and can include the following:

- Advertisements
- Recruiting fees
- Sign-on bonuses
- Salary (a salary increase may be necessary to attract and retain qualified applicants)
- Benefit costs

Indirect costs can include the following:

- Time invested in employee search and interviews
- Loss of productivity
- Reduction in product or service quality

Indirect costs are generally difficult to quantify and is therefore often omitted as a cost element. For example, consider that trained employees will take their knowledge and skills with them when they leave the company.

The organization will incur a loss of productivity or increase the potential for mistakes by new employees until they become fully trained to perform the work. Often, calculating these costs may not be considered or easily quantifiable. Therefore, when Lean impacts workers to the point that they believe it to be necessary to seek jobs elsewhere, a cost to the company has incurred. This is a classic time when talent management comes into play. A good talent management plan should be a part of the overall corporate strategic plan. The plan can save a company significant revenue expenditures that can be reinvested into the business or transferred to its stakeholders.

6.2 What Is Talent Management?

Talent management is the means used by an organization to recruit, develop, and retain the talented employees needed to ensure that the organization is able to meet its mission and deliver what is expected by stakeholders and customers. It is not a plan that is invoked when workers are talking about leaving the company. This plan must be a living strategy that is actively used and viewed as active and effective by the workers.

In the work environment, it is a constant struggle to attract and maintain qualified resources especially in times when the economy is flourishing. A changing work environment adds complexities and can add difficulty in retaining workers. The impact of Lean implementation is yet another factor that must be considered and handled appropriately to avoid a loss of talented workers fearing change and potential job loss. It is more economical to work at keeping employees than to allow them to walk out the door and spend money on recruiting and training new workers that will need time to get up to speed.

6.3 Balancing Employee Needs with Organizational Demands

Talent management and retention is an important exercise in balancing the needs of the company and the needs of the employees. Recognizing that managing and retaining the resources needed to render an organization successful takes financial investment. The question is how much financial investment is necessary and practical? Organizations across the board are constantly weighing the needs of the business with the needs of employees. It is an act that is difficult to balance. Some of the needs of the company and employees that must be considered when developing a retention strategy are listed in Figure 6.1.

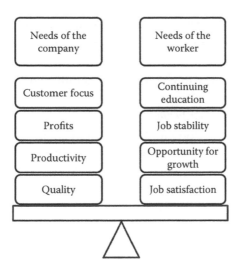

FIGURE 6.1
Balancing needs.

An important goal when balancing needs is to provide the accommodations that employees will value that can help form a bond with the organization increasing retention potential.

6.4 The Value of a Flexible Workforce

A flexible workforce is being defined as a group of employees who are capable of performing a variety of functions within an organization. Flexibility in the workplace is an essential attribute for employees and employers. Flexible workers generally have more diverse skills than workers who are less flexible. These workers are often called upon to take on a diversity of challenging assignments within the organization, which results in an expanded knowledge base and visibility with the leadership team and often leads to another position of authority or advancement opportunities.

A streamlined workforce, rapid growth in technology, and a changing market are just a few reasons why flexibility is rapidly becoming the new norm for many organizations. A flexible workforce is optimal for implementing Lean and adapting to changes as they occur within an organization. Advantages of flexibility within an organization are as follows:

- *Advantages of employee flexibility*
  - Embraces change easily—today's work is fluid and constantly changing. Employees who are able to adapt to shifting priorities and changes in the way business is being conducted are a must.

- Lean becomes easier to implement and sustain.
- Embracing changes can expand opportunities and open doors to new adventures.
- Aid in the ability to obtain and maintain a sustainable work–life balance.
- *Advantages of employer flexibility*
  - Being flexible is good for business and creates a win–win proposition between workers and the leadership team.
  - Being flexible with your employees builds employee trust and commitment to the leadership team and the organization.
  - It also helps attract and keep talented workers and spark creativity in team members to engage in helping find solutions to address issues.
  - Flexibility demonstrates that diversity in the workplace is valued by the management team.

6.5 Knowledge Retention and Transfer

Many organizations are concerned about losing the knowledge and expertise of their employees. These concerns have become even more real for companies that are dealing with growing retirement trends and frequent downsizing resulting from a change in business processes, practices, or technology resulting from implementing Lean thinking, mergers, and acquisitions. Loss of knowledge can expose a business to loss of productivity, increase injury and illness among the workers, regulatory noncompliances, and the list goes on.

Knowledge retention and transfer involves capturing knowledge in the organization so that it can be transferred to another member of the organization and used at a later date. It is critical that knowledge retention be integrated into the way a company operates and begin well before a key employee is preparing to depart the company. A good knowledge retention strategy will assist a company in being ready to ensure that the staff needed will be available and answers the critical questions shown in Figure 6.2. Knowledge retention is not feasible if the knowledge that is required and the associated consequences are unknown.

One can approach knowledge retention and transfer from different perspectives. Some knowledge preservation techniques can include the following:

- Implementing a technique that involves documentation of the job content by workers

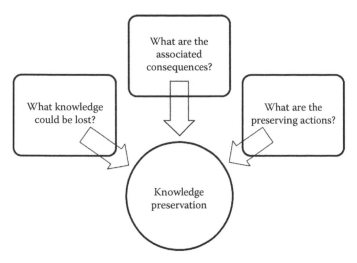

FIGURE 6.2
Knowledge preservation questions.

- Identifying employees who can be proficient in more than one job
- Identifying key positions and performers who could be developed for those roles

6.6 Employee Retention Strategy

When speaking of employee retention, reference is being made to an organization's ability to hold on to his or her talented resources. To do so, there needs to be a focus on what policies and practices are necessary to retain employees. The various policies and practices will form the basis of the strategy used for retention. Every organization invests money in training new employees to prepare them to assimilate into the organization. A good practice would be to invest adequate funds on retaining employees and to implement a good retention strategy. This action can potentially limit the amount of employees needed to be assimilated into the organization to replace workers who are leaving. Employee retention takes into account and focuses on the various measures that can be taken so that employees stay in an organization for the maximum possible period of time. There are some employees whom a retention strategy will not impact as these employees may have a career strategy of spending limited time in the same job, company, or location. A good strategy should take into consideration the attributes listed in Table 6.1.

TABLE 6.1

Elements of a Good Retention Strategy

Element	Contribution
Retrain or replace poor leaders	Poor leaders can contribute significantly to worker's unhappiness with the organization and productivity loss. Poor leaders dampen employee morale and employee engagement.
Recognize the contribution of employees	Employee recognition adds to employee morale and communicates to the employee that his or her contribution is recognized and important.
Encourage open and honest communication	Facilitate transparency through open and honest communication. Communicate frequently and openly on how the business is progressing and the contribution of employees to business outcomes.
Incorporate technology where feasible	Technology can be used to help employees keep current in their field of practice and connection with colleagues across the globe.
Provide technical and leadership training	Training provides employees confidence that they are prepared to perform their job and demonstrates that management is willing to invest in them.

Losing colleagues can have an impact on the rest of the team and the organization. Other workers may feel uneasy and stressed if they see talented workers leaving the company. Many leaders do not have a plan to retain their workers, or if a plan is in place, it is generally not effective. One may ask, when is the best time to focus on employee retention? The answer to that question is that staff retention is a constant activity that often involves engaging one's creative abilities. The scenario in Vignette 6.1 represents a poor way to address retention of workers, yet some managers today are still using this philosophy.

VIGNETTE 6.1

Generally when an employee communicates to management his or her intension to depart the company, he or she has already accepted employment with another company.

An employee retention program is not a practice of promoting a key resource before he or she walks out of the door or providing a retention bonus to stay with the company. Many companies use the two aforementioned methods as a substitute for a retention plan to retain critical staff. This action is a precedence that is not worth setting.

Retention of talent is key to continued growth and success of businesses that it is worth investing the funds, time, and effort to ensure that these individuals are satisfied enough to stay and develop within the company instead of looking elsewhere for professional opportunities and career growth.

Talented employees contribute to the enhancement of business in several ways such as

- Ensuring customer satisfaction
- Maintaining productivity
- Driving product and system development and innovation

Retaining employees, especially the ones who are engaged and dedicated to the organization, requires a strategic and sensitive approach to the way people integrate into the work environment. Managers must not underestimate the importance of retaining individuals who possess the skills and are pertinent to achieve business success.

6.6.1 Employee Recognition

Employee recognition is an effective communication tool that rewards people for contributing above and beyond to important business outcomes. When people are recognized, behaviors and actions are being reinforced that you would like to see repeated. An effective employee recognition program should be simple and immediate. Three powerful benefits of an effective employee recognition program include the elements listed in Figure 6.3 and are described as follows:

1. *Boosts confidence*: Recognizing employees for their work and achievements generates a feeling of pride and fulfillment in knowing that they are doing a good job and contributing to the success of the organization. Everyone enjoys being praised by another, especially someone who is in a position of authority such as a supervisor or a manager. The feeling of pride and excitement generated through recognition creates confidence that can directly improve and sustain performance.

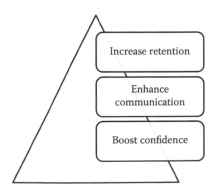

FIGURE 6.3
Powerful benefits of employee recognition.

2. *Enhanced communication*: Communicating with workers about their exceptional contributions to the company opens up the communication channels. It makes management appear more approachable. Recognizing employees show the softer side of the management team. This is a side that may be infrequently visible to the staff because management often interacts with workers only when providing directions of instructions on getting work done.

3. *Increased retention rate*: An effective employee recognition program can improve your retention rate for employees as well as help create a positive work culture. When employees work in an environment where they are not recognized and feel like they are not appreciated and valued, the risk of them walking away to work for another organization increases.

Recognition has consistently been a strong motivator in the work environment and can range from a monetary award, a trinket, or even just saying the words *thank you great job*. Recognizing employees highlights and reinforces the specific behaviors or actions that are key and valued. Admittedly, it is not always easy to maintain an effective recognition program. However, a good program can pay for itself many times over.

6.7 Talent Care

Developing employees takes time, money, and, at times, patience. Therefore, a candidate who has been selected as a talent that the company would like to retain should be provided with every incentive feasible so that he or she will want to remain with the company. This means that he or she should not be ignored or treated unfairly by management and the company. The human resources policies and practices should be supportive of the succession planning process. The action of management is critical in caring for and developing the candidate. Talent care is even more important when changes are being made within a company. When workers feel that they are not important and are being disregarded by management and the company, they tend to lose any allegiance to doing what is in the best interest of the company. At this point, the *it's all about me* syndrome takes over. The scenario in Vignette 6.2 is a real-life example of what can and often does happen.

VIGNETTE 6.2

While traveling with an airline that recently merged with another, a manager experiences the raft of unhappy disgruntled employees. During his or her travel, he or she went through a couple of airports interacting with employees of the airline. In several cases, the employees were not helpful and accommodating as one would expect of someone who worked in the customer service industry. It was clear that the workers were disgruntled because of the changes that were made as a result of the merger. In fact, at the two airports, the workers began confiding in the manager, a complete stranger, their unhappiness with their employer. The employees communicated to the manager that the company was trying to get rid of their jobs. As a result of the manager's experience with the employees over a couple of days, they are now reluctant to choose to fly with that airline as a preferred carrier for future travel needs.

6.8 Why Do Employees Leave Their Companies?

Many researchers and practitioners believe that employees leave their organizations primarily because of frustration and poor relationship with their management. Other reasons why employees leave their organizations are shown in Figure 6.4 and are discussed in more detail in Sections 6.8.1 through 6.8.5. It is management's responsibility to make every attempt to provide a work environment that will stimulate workers in a way that they

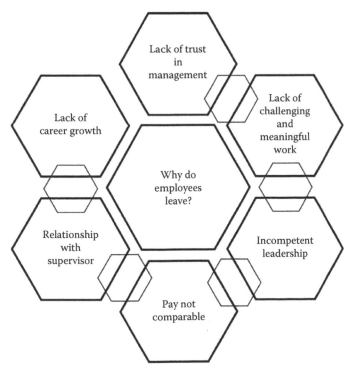

FIGURE 6.4
Why do employees leave their employer?

would want to remain with the company and to provide them with meaningful and challenging work.

6.8.1 Relationship with the Supervisor

Employees do not need to be good friends with their supervisor; however, they need to have a professional relationship that is based on trust and respect. The supervisor is an integral part of a worker's daily life while at work, often more than eight hours per workday. An uncomfortable relationship is not productive and can be stressful for both parties. A bad relationship between the supervisor and worker can impact employee engagement, commitment to the organization, and confidence in the leadership team.

6.8.2 Incompetent Leadership

Leaders do not need to know all of the technical aspects of a job in order to be considered competent. However, they should be skilled in the attributes of a leader. Mastering the fundamental skills of leadership renders a leader competent in the eyes of others. An incompetent leader can

- Contribute to an employee making the decision to quit his or her job
- Impact negatively organizational trust
- Cause an employee to withdraw and limit engagement in work activities
- Reduce the effectiveness and efficiency of the organization

An incompetent leader can cause a company great harm that can significantly impact the bottom line. The impact of an incompetent leader is difficult to quantify when calculating the bottom line. We have talked about what is necessary to prepare the workers for process changes; however, management must also be prepared for change. Before implementing Lean, it is a good idea to ensure that the leadership team is competent and onboard to lead the charge and set the example of change acceptance and implementation. A project, process, and even a company can fail if managers do not have the knowledge and ability to lead competently.

6.8.3 Lack of Challenging and Meaningful Work

For the most part, workers want to enjoy their work and be engaged while at work. No one wants to be bored or not challenged by his or her work. Employees who are engaged in their work are willing to perform in a way that will help an organization to achieve its mission. Not only must work be challenging for workers, it must also be meaningful. Most people want to do something that is meaningful, something that will make a difference, and something that will make them feel good about themselves and their existence. It is incumbent upon managers to provide work that contributes to the mission and help employees understand the value and the importance of their contribution. If management is not actively seeking ways to help workers connect with their job and feel like they contribute, they will find an employer who will provide the satisfaction they seek. If workers are not challenged and engaged at work and are not allowed to use their skills, there is a strong possibility that they will not remain with the company.

6.8.4 Lack of Career Growth and Advancement Opportunities

Having access to career growth and development opportunities is a desire of most employees. They want to continue to develop and grow their skills. They would also like to have opportunities to function in a variety of roles and advance to higher levels within the organization. If they are not able to grow, develop, and be promoted at their current company or organization, they will seek employment elsewhere leaving companies with gaps to fill. The benefit of advancement opportunities in a company is a significant contributor to employee retention. A retention strategy is ineffective without the opportunity for career growth.

6.8.5 Comparable Pay Does Matter

Many times I have heard managers say that *pay for work performed* is not as important to workers as the satisfaction experienced from the work itself. Today, workers are concerned more and more about pay to the extent that they are being compensated in line with their peers. Some workers view pay as an opportunity for managers to show that they are valued. Workers will leave their company when they believe that they are not being compensated equitably for the work performed in comparison to their peers internal and external to the company. Today it is easier for workers to have access to market information that provides a general compensation pay range of what workers are being paid to perform the same or similar jobs external to the company. Most companies perform a market analysis at some frequency to determine if they are paying their employees comparable to the market. This is a good practice to aid in ensuring equitable pay for the work performed. Pay does matter and it can make the difference whether employees stay or go.

6.9 Why Do Employees Stay with Their Organization?

There are many reasons why an employee may choose not to stay with his or her organization. Many of these reasons are shown in Figure 6.5. Designing an employee retention program that will address the elements listed in this

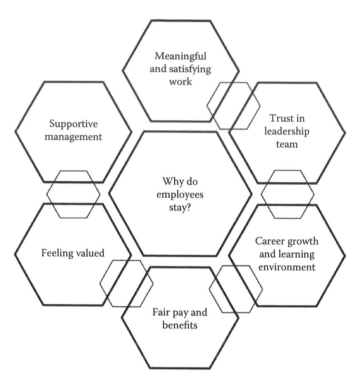

FIGURE 6.5
Why do employees stay with employers?

figure can go a long way toward helping retain workers and reduce the company payout when it comes to having the ability to attract and retain the skills needed to perform work and continuously deliver value to its customers. These elements will be discussed in more detail in Sections 6.9.1 through 6.9.6.

6.9.1 The Impact of Meaningful and Satisfying Work

Meaningful and satisfying work provides the stimulus for people to go to work day after day and exert the effort needed to succeed. Meaningful work can be a good predictor of worker's job satisfaction and absenteeism. Employees feel a sense of pride, accomplishment, and confidence when they are able to use their skills and abilities in the workplace. This feeling of accomplishment fuels the desire to go the extra mile and do more. Management must always keep at the forefront that workers have a desire to grow, learn, and progress. For workers, meaningful work has the following components:

- Work that requires them to think and use their knowledge and skills.
- Something that they enjoy doing.
- The meaning and value of the work are important to the company's mission.

- Complements their skills and abilities.
- Sparks their creative thinking.

When workers find meaning in the work they perform, they

- Are happier and motivated.
- Are more satisfied at work.
- Engage in workplace activities.
- Tend to remain with their current employer for a longer period.
- Show high individual performance.
- Feel personal fulfillment.

6.9.2 Trust in the Leadership Team and Its Impact

More and more trust is being recognized as a critical element of organizational success and increase in productivity. When employees trust their leadership team, they are willing to do whatever they can to ensure success of the business. Employee trust in the leadership team can facilitate the following:

- Employee retention
- Innovation and creative thinking among workers
- Productivity
- Employee focus on performing task safely

When employees trust their leaders, they are more subjected to authority and vested in the success of the mission of the organization, which can serve as a means to provide the encouragement needed by employees to remain with the company.

6.9.3 Career Growth in a Learning Environment

We know that employees do not like to become stagnant in their careers and would like to feel that they are progressing. An organization that can offer advancement and growth opportunities to its employees is more likely to retain them for the long haul. Workers are more likely to be more engaged in workplace activities because they can see the possibilities of future. In organizations that have opportunities for career growth, employees are more engaged and are able to deal with change easier. In fact, career growth and the benefits of a learning environment are two important elements that contribute to employee satisfaction and retention.

6.9.4 Fair Pay and Benefits

We have been hearing a lot about fair pay for workers across all industries. Most of the talk has been associated with executive type positions. However, fair pay is an issue at all levels, across all job functions, and in all industries. When speaking to employees, they will tell you that comparable and fair pay does matter. Therefore, it is incumbent upon managers to design a process to ensure fair and equitable pay for everyone. This does not mean that high performers cannot be compensated at a higher rate. The point is that compensation policies and practices must ensure that employees are treated fairly across the board. In addition, the company's overall benefit plan must be comparable to industry standards. The compensation and benefit portfolio can be a significant factor in the retention of employees.

6.9.5 Feeling of Value

Everyone wants to feel valued for his or her contribution in relationships as well as work. Employees who do not feel that they are valued at work will most often become distracted and can become less productive. Employees who feel valued for the work they performed are

- Generally motivated to do their jobs.
- Willing to continue to improve performance.
- Having less absenteeism.
- Willing to be engaged in the workplace.

6.9.6 Supportive Management

For employees, the importance of a supportive leader team cannot be underestimated or undervalued. Most managers will verbally provide support to their workers; however, their actions are silent or counter to what is being verbalized. This lack of action can leave employees seeking something else from someone else. There are fundamental actions that employees expect from management to demonstrate that he or she support them. These actions include the following:

1. Supporting their career objectives
2. Providing training opportunities
3. Providing a positive work environment
4. Providing fair recognition and rewards program
5. Providing advancement opportunities

Management must be careful not to allow his or her own agenda to collide with his or her role as a leader. He or she must ensure that he or she is able to make sound and subjective decision with respect to supporting the growth of each worker. If an employee believes that his or her management is not acting on his or her behalf, he or she will not allow himself or herself to trust or become committed to the organization. The scenario in Vignette 6.3 is a real-life event and demonstrates actions of a manager lacking in some fundamental skills. These actions can make employee retention difficult, to say the least.

VIGNETTE 6.3

Consider the character of a manager who continues to communicate that a top performer is not ready to advance to the next level, which will make this individual a peer. The top performer was moved to another department and performed extremely well. Soon the top performer excelled above the manager who attempted to keep him or her from advancing.

The scenario depicted in Vignette 6.3 led to employee distrusting the management and providing skepticism for every word that the manager spoke and every action performed by that manager. In addition, because words can travel, the employees shared their experience with others who subsequently were not willing to work for the respective manager. The point is that the support and actions of managers can determine what actions employees will take on their own behalf.

6.10 Summary

Talent management and retention is not an easy task for managers. Although the task can be challenging, it must be addressed and considered as an integral part of a company's operating strategy. In this chapter, many ideas and suggestions were introduced as to why workers are unhappy or why they may even decide to seek employment elsewhere. The ideas presented are not all inclusive; however, they do provide the basis and the importance for the need to plan ways that will encourage workers to remain with the organization.

Managing a skillful workforce is challenging partly because of the competition faced from other employers and the diversity of choices available to the skilled worker. One may enquire as to why this is important because the misconception of some leaders is that workers are paid for a service and that all they are obligated to provide. Talented and skillful workers are not always easy to recruit and hire. Therefore, it is incumbent upon leaders and managers to pay attention to the needs of the workers who go beyond a paycheck.

Talent management can emerge as a significant issue when Lean is being implemented. Workers tend to see uncertainty when things change that may impact them. Lean implementation is a fundamental change in the way business is being conducted, and Lean cannot, and will not, be successful without the skillful workforce needed for implementing and maintaining the process. Thus, some workers may see this as a sign that the company is seeking to optimize its profit through reduction in the workforce. Talent management is a big job, but remember that in doing so, a little goes a long way, and a stable workforce is important when implementing process improvement initiatives. Thus, employee retention and knowledge retention are beneficial in charting the path toward success. Paying close attention to the factors listed in this chapter can help reduce turnover and help retain critical skilled employees. If not, be prepared for holding regular employee search activities, interviews, and onboarding of new workers.

7

Employee Development and the Hidden Cost

7.1 Introduction

Training, knowledge development and retention, and career development are very important for a company or an organization that is seeking to progress. Training is defined as the process of acquiring the essential skills required to perform a job successfully. Effective training is designed to target specific goals such as an individual's understanding of a process, practice, procedure, or technology that is required to complete work safely and effectively.

Training that is designed for career development places emphasis on skills that are applicable to and required, in order to adequately work through a wide variety of situations and issues. This type of training usually focuses on areas such as leadership, decision making, critical thinking, communication, and leading people. Developing and training employees take time, money, and, at times, patience. It is not a one-time act but a continuous process. Developing and administering training represent an added cost that can result from many sources. Some of these sources are listed as follows:

- Time and money used to procure material that will be used in the training
- Training development time
- Securing an instructor or company to administer the training
- The time employees are in class being trained and not performing their assigned jobs
- On-the-job training (OJT)
- The cost of retraining if necessary

The list above represents some of the cost-producing elements that can be associated with training that are not always quantified and not included in any funding model. Depending on the process or technology being implemented or changed, there may be other cost-producing elements to consider outside of the list provided. Many times, Lean implementation results in the need to train workers on a new process or technology incurring cost that may not have been anticipated.

7.2 Why Train?

Training is important for business to ensure that the workers delivering services to and on behalf of the company have knowledge in accomplishing the task safely and efficiently. There are many laws that require employers to train employees. Complying with these laws can help reduce employer's liability. Trained employees are also the key in ensuring that the product delivered to the customer is of high quality. Before embarking on the training journey, there are a couple of questions that should be asked to ensure that the training being contemplated is the right training to develop or procure. These questions include the following:

1. What is the purpose of the training?
2. What knowledge is deficient?
3. What is the target population?

Answering some fundamental questions can highlight when and why training is required. Figure 7.1 provides some fundamental benefits of why training may be necessary and the benefits that can be realized. A brief summary of each benefit is discussed:

1. *Addressing weaknesses in employee skills*: Some workers have weaknesses that must be addressed through training. If not addressed, these weaknesses can stifle workplace performance and employee growth. Training assists in eliminating weaknesses through strengthening worker skills and knowledge. The knowledge gained through training produces workforce reliability in getting the work done with accuracy.

2. *Improve worker performance*: A properly trained worker has the knowledge and confidence to perform his or her job responsibilities with ease. He or she is likely to use his or her knowledge to ensure that work tasks are performed to ensure that the goals of the company are met.

3. *Increase productivity*: Training also provides a means to increase productivity because workers have the knowledge and skills needed to perform routine tasks in the same manner. Trained workers can generally perform at a faster rate and with efficiency and accuracy, thereby increasing productivity for the task.

4. *Provide consistency in performing a task*: Workers receiving training with the same objective and content are expected to perform their job responsibilities in a consistent matter. This consistency in performing work should reduce the amount of mistakes or rework needed to deliver quality products.

5. *Improved quality level for products produced or services provided*: Standardized training provides the means for workers to implement

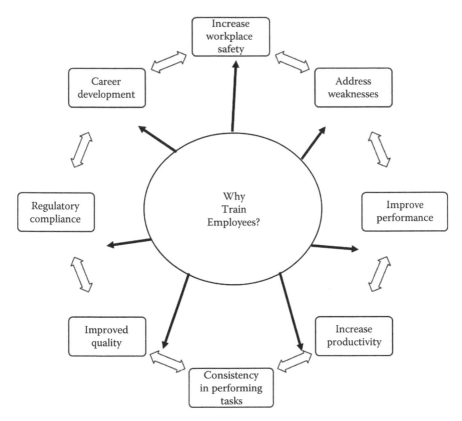

FIGURE 7.1
Benefits of a trained workforce.

standard methods of performing tasks providing an opportunity for uniformity of products and services.

6. *Improve worker safety*: Training that focuses on hazard recognition and mitigation is essential in helping workers identify and reduce workplace hazards. Workers with the appropriate level of knowledge in injury and illness prevention, hazard recognition, and mitigation can be instrumental in helping management plan and carry out work safely.

7. *Regulatory compliance*: There are many federal laws that require employers to train employees on specific topics. In such cases, training may be tied to a company's ability to demonstrate compliance, and having the appropriate training administered to workers can reduce the company's risk and liability.

8. *Career development*: This is a great way to demonstrate value for the employee and to ensure that the company is able to cultivate and retain the knowledge needed to sustain the company.

A good resource to use when developing training is outlined in Vignette 7.1. The model discussed in this manual represents the most popular training methodology followed and endorsed by members of the training community.

VIGNETTE 7.1

Many experienced trainers follow a methodology to training that is known as ADDIE. This method is a five-phase approach. The acronym means:

A, analysis; D, design; D, development; I, implementation; and E, evaluation.

The *DOE Training Program Handbook* presents a systematic and comprehensive approach to training: http://www.publicpower.org/files/PDFs/DOEHand book TrainingProgramSystematicApproach.pdf.

7.3 The Role of the Leadership Team

The leadership team is responsible for fostering a learning culture within the company. Developing employees should be a top priority for supervisors and managers. This includes identifying the competencies that an employee will need to perform effectively in the position he or she has been hired to perform. Knowing the required competency is necessary to build a qualified and highly effective workforce. Leaders who demonstrate their commitment to advancing the knowledge and competency of the worker can create an organization of loyal, committed, and motivated workers.

The leaders of a company are responsible for supporting training and ensuring that the needed resources to include funding are available to support training development and delivery. When management is not supportive and engaged in ensuring that training is available for members of the staff, training may be limited, and work and product quality may not be complementary to the needs of the organization.

7.4 Training Needs Analysis

A training needs assessment is used to identify whether training is the appropriate action to take to correct a specific performance problem. The fundamental reason for conducting a training needs assessment is to identify the needed performance requirements and the knowledge and skills required by employees to accomplish work to meet the goals of the company. The assessment is used to determine the gap between the knowledge and skills that organizational members currently possess and what is needed to meet the goals and objective of the organization. The analysis should also explore and document the consequences for not addressing any gaps identified. A good training needs analysis should answer at a minimum the following questions:

1. Why is training necessary?
2. What knowledge is deficient?
3. Who needs to be trained?
4. When will the skills be needed?

Training needs analysis can be bundled into three categories. These categories are organizational, task, and individual. Each of these categories is discussed in Table 7.1. Tables 7.2 through 7.4 cover potential questions associated with each category that should be considered.

TABLE 7.1

Training Needs Analysis Categories

Categories	Benefits
Organizational	• Used to align training with the strategy of the organization. • Used to align resources and management support for the training.
Task	• Important in identifying the task necessary to perform work and successfully integrate into the organization culture. • Seek to identify the knowledge, skills, and behaviors as they relate to consistency in performance.
Individual	• Used to determine how well an employee is performing a task and determines the individual's capacity to do new or different tasks. Individual assessment is effective in providing information on which employees are in need of training and what type of training is needed?

TABLE 7.2

Organizational Assessment—Sample Questions

Organizational assessment: The basis of this assessment is to analyze the effectiveness of the entire organization and identify any issues that can impact organizational effectiveness. When effective, the assessment is used to reveal the competencies, knowledge, and skills that are needed by an organization to bridge all gaps within all groups or department within the organization.

Questions:
1. What are the tasks that are essential for company's success?
2. What are the skills needed to ensure success of the company's mission?
3. Where training is needed in the organization?
4. What are the gaps in training?
5. Will training be effective upon implementation?
6. How will training be conducted?
7. When will training be conducted?

TABLE 7.3

Task Analysis—Sample Questions

Task assessment: The task assessment focuses on information concerning a specific job
function. When effective, analysis identifies the competencies and skills required to perform
the job effectively and efficiently. Task analysis is used to discover specific training needs, as
it relates to the various tasks to be performed within an organization.

Task Description _____

Questions:
 1. Is this task required to complete essential work?
 2. Is the training required by a federal, state, or local regulation? If so, list regulatory
    citation.
 3. How often is the task performed?
 4. Is the task critical to performing the job? Provide an explanation of the response?
 5. What behaviors are necessary to successfully perform task? What behaviors are critical?
 6. Can this task be effectively trained on the job? Provide an explanation of response.
 7. What prerequisite knowledge and skills are needed?
 8. Are there any consequences if the task is not performed or not performed correctly?
    Provide an explanation of the response.
 9. What level of proficiency is expected after training?
10. How will task proficiency be measured?

TABLE 7.4

Individual Assessment—Sample Questions

Individual assessment: The individual assessment focuses on a particular employee to
determine how well he or she is performing and identify gaps in knowledge and
performance. This type of assessment, when effective, determines the individual's existing
skills and competencies, and whether he or she is capable of performing a specific task or job.
The individual analysis identifies who within the organization is in need of training, and
what type of training is needed. The individual assessment forms the basis for the
development of a training and development plan for the individual employee.

Name: _____

Job Tile _____

Task being Evaluated _____

Questions:
 1. Does the individual have the skills needed?
 2. Does the individual have the capacity to learn the skills needed?
 3. Does the employee have the willingness to be trained?
 4. What are the skills and knowledge needed to effectively perform the task?
 5. What skills are deficient?
 6. How much training is required to close the gap?
 7. How much time required to close the gap?
 8. Who will perform the task until the individual is able to perform task?
 9. How will training be administered?

The assessment should be conducted early and before allocating any budget to address training needs and developing training. A training needs analysis is necessary when a change in process, technology, or implementation
of Lean process improvement initiatives. When a needs assessment is not

conducted, a training may fail to address the performance needs or knowledge gaps, and therefore it can result in unanticipated or potentially unrecognized cost. The importance and impact of conducting a needs assessment must not be underestimated in terms of its importance. The assessment, if effective, is the key in addressing the resource needs required to fulfill the mission of the company, improve productivity, and consistently provide quality products and services for the customer.

7.5 Training Delivery Methods

There are several methods that can be used to deliver quality training that is effective. Often, when we speak of training, one would generally defer to formal classroom training. There are times when classroom training is appropriate as a stand-alone method or as long as it is supplemented with other training methods such as OJT, mentoring, coaching, or other means of training (Figure 7.2). Although some of these methods may not be considered traditional training, they do offer a means to get a worker up to speed on the task and increase their knowledge. We will touch upon each of these methods briefly in Sections 7.5.1 through 7.5.6. All or some of these methods are appropriate to fill the gap in knowledge after implementation of Lean or other changes in process. Regardless of which training method or methods used, a cost is associated with delivering the training.

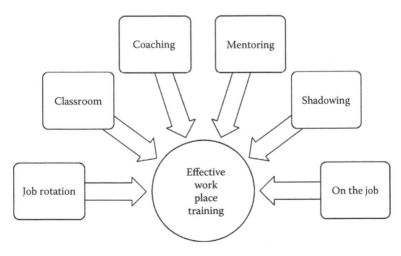

FIGURE 7.2
Workplace training methods.

7.5.1 Classroom Training

Conducting classroom training is a long-standing traditional way of delivering training. Therefore, it is the most common method used to develop training. Different workers have different preferred learning styles, and many still have as a preference the face-to-face classroom-training experience. In the classroom, there is an opportunity to collaborate and solve problems with other workers. To be effective, classroom training should have the means to evaluate employee learning. Some advantages and disadvantages of classroom training are listed in the following. These advantages and disadvantages are important to consider when deciding whether to use classroom training to close the gap in knowledge.

The advantages to classroom training include the following:

- Training can be conducted for a small or large group of workers at one time.
- Group interaction can enhance learning since employees have an opportunity to learn from each other as well as from an instructor.
- Classroom learning allows employees to interact with a trainer in a way that is not feasible with training administered through electronic means such as online.
- It allows employees the opportunity to learn in a safe and quiet environment free of noise and distractions.

Some disadvantages of classroom training include the following:

- Employees are taken off the job for training, which can create the need for overtime to get work accomplished and impact production schedules.
- It is often difficult to arrange classroom training for people who are working on the night shift.
- Classroom training often lacks the ability to introduce hands-on experience.

7.5.2 On-the-Job Training

Hands-on training or OJT is most effective when performed in the workplace under actual work conditions. OJT is designed to teach the skills, competencies, and knowledge that are needed for employees to perform their workplace tasks. If training is not performed on the actual job site, the training location should closely resemble the actual job site. OJT uses the existing workplace procedures, equipment, knowledge, and skills to help employees learn how to perform their job effectively. Training occurs within the working environment in which an employee is expected to perform his or her daily tasks. Advantages and disadvantages of this method of training are listed as follows:

Advantages

- Often, more cost effective than other type of training methods.
- No outside trainers are necessary.
- Training is conducted during employee work shift.
- A new worker has the opportunity to become acquainted with coworkers.
- Can provide immediate feedback on performance.

Disadvantages

- OJT can be time consuming for the trainee as well as the supervisor or experience employee conducting the training.
- Planning and evaluation can consume time of the trainer.
- It may be a challenge to find the right person to conduct the training. The person conducting the training must be knowledgeable in all aspects of the job and must have the ability to teach effectively.
- Can disrupt the flow of work.

7.5.3 Mentoring

Mentoring involves pairing an employee with an experienced coach to oversee his or her learning and growth. The mentor provides advice, instruction, and guidance on job performance and other work-related activities. The mentee learns a job task and consults with the mentor for assistance, answering questions, or clarifying any discrepancies.

Some advantages of mentoring include the following:

- One-on-one training is personalized for the mentee.
- The mentor is available to offer advice as needed.
- Easier to facilitate open dialogue.

Mentoring is a tool that many organizations use to grow their workforce on a continual basis. An employee can benefit from a mentoring relationship because he or she has someone with more knowledge and experience to consult with as needed. During the early stages of an employee's career, a mentor can provide tips on career growth and introduction to other professionals, and as the employee matures in his or her career, a mentor can continue to be a valuable advisor. Some employees may need a network of mentors to build

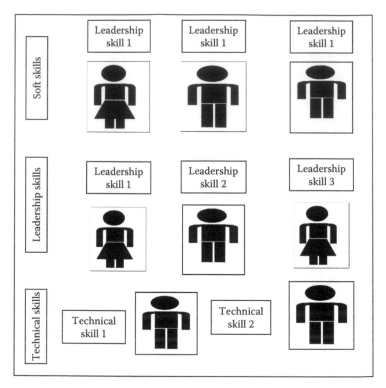

FIGURE 7.3
Mentoring network.

the skills necessary to successfully complete their job responsibility; in such case, mentoring can become too intensive and expensive. An example of a complex mentoring network is shown in Figure 7.3.

7.5.4 Job Rotation

Job rotation is an effective way to teach workers how to become proficient in performing various tasks. This form of training allows employees to rotate to different jobs within the organization performing and becoming proficient in unrelated tasks. There are some distinct advantages and disadvantages of job rotation that must be considered.

The advantages of this method of training are as follows:

- Reduces the potential for employees to become bored performing the same task day after day
- Increases skills and knowledge in more than one task
- Opens new opportunities for employees to advance
- Facilitates flexibility in assigning work

The disadvantages of this method of training are as follows:

- Employee may not be satisfied with all of the tasks in the rotation.
- Can be seen as disruptive to the workflow until employee becomes proficient in the tasks included in the rotation.

7.5.5 Coaching

Coaching is an effective way to develop workers' skills, knowledge, and abilities, and to enhance performance. It focuses on concrete issues such as thinking strategically, managing more effectively, or development in other specific competencies. In order for coaching to be effective and to achieve the intended result, an effective coach must be onboard. Coaching should be considered when

- An organization is seeking to use performance management to develop employees in specific competencies.
- There is a change in current process or system or introduction of new process or system.
- An organization has employees who are not meeting expectations.
- An employee requires additional skills to complete his or her job successfully.

Effective coaching can equip workers with the knowledge and skills that can provide an opportunity for self-development, increase effectiveness in the work they perform, and increase their value to the company. Coaching can be and is often used to compliment training and development activities such as after completion of a training course to help workers implement what they have learned. Some disadvantages and advantages of coaching are listed as follows:

Advantages

- Can cost the organization less than other traditional form of training since the coach is already on the payroll
- An effective means to grow employee in the work environment

Disadvantages

- Finding the right coach is not always feasible.
- An individual may not connect with the assigned coach.
- Coaches may have limited time due to their schedules to prepare for and conduct sessions.

7.5.6 Shadowing

Job shadowing is an effective OJT method that has been used effectively to train new or existing employees by affording them time to observe other trained employees perform the task that they will be expected to perform.

Some of the advantages of job shadowing include the following:

- Represents a lower yet effective cost effort to train
- Develops skills effectively
- Fosters and develops team spirit between employees
- Builds employee confidence

Some disadvantages of job shadowing include the following:

- May be difficult to schedule.
- Long-term experience employees may provide an inaccurate way of performing a task.
- Employee being shadowed may become concerned about the stability of his or her role with the company.

7.6 Training Constraints in Lean Environments

In Section 7.5, we discussed several training methods that have proven to be effective in training workers to effectively and safely perform their tasks. Although these methods are effective, there are challenges with implementing these methods with ease when working in a Lean environment. Traditionally, in a true Lean work environment, resources are streamlined with little to no excess bandwidth. This lack of bandwidth can create complexities in any attempt to carry out a clearly defined training strategy. Therefore, the training strategy has to be well thought out and capable of being implemented with some level of flexibility in order to provide value for the company and the workers.

Although training may represent addition challenges for Lean work environment, it can be accomplished through extensive creativity, planning, and strategizing on ways to ensure workers receive the required training. Also, there may be a need to manage workers' expectation on the level and amount of funds available for training, certification maintenance, as well as the ability for the organization to support achieving new certifications in the professional field. Table 7.5 lists some issues that should be considered when developing and implementing a training strategy.

TABLE 7.5

Implementing Training in a Lean Environment

Training Methods	Implementing Difficulty
Job rotation	The organization must have the bandwidth to allow employees to move into other roles for the purpose of training. During the time of the training, productivity may be hampered and reduced because employees will be in a training role as opposed to contributing completely to the task. In addition, the job that was being performed by the employee in training to learn new skills has to be performed by someone else. Job rotation can facilitate the need to spend more money to get the work accomplished and to meet deadlines. For example, through overtime by current employees or through the hire of temporary workers to complete the work.
Coaching	Because coaching is generally a short-term activity that is focused on one discrete activity or behavior, it can be a viable option to address issues or behaviors that can be rectified in a short-time frame. This is a good training option for Lean environment because the investment is minimal.
Mentoring	Mentoring is relationship based and is a long-term relationship between the mentor and mentee. Therefore, it requires the commitment of both parties. However, it generally takes the least amount of time in that sessions can be spaced out across a long period of time using short sessions. It is the least disruptive and cost to an organization.
Classroom	The challenge with classroom training is that the worker is taken away from his or her daily jobs to a setting to learn new concepts. Work must be carefully planned to allow workers to attend training and not significantly impact production and the schedule.
Shadowing	Job shadowing represents a potential low-cost way of using current skilled workers to transfer knowledge to new or existing employees. It does require the ability to allow employees time to shadow and for employees to train. This method can slow down productivity and has the risk of employees not being trained adequately.
On the job	On-the-job-training possibly represents the most effective means of training to transfer knowledge to workers. However, it can be extremely time consuming and relatively costly, because some candidates may require extensive training. The staffing level may make this option impossible for some Lean organization.

7.7 Continuing Education

The workplace and technology are constantly changing as such continuing education is a necessity for workers to remain current in knowledge, skills, and changes in technology related to their field of expertise. Many professions require continuing education to comply with laws and to remain licensed or certified. When leaders invest in continuing educational opportunities for employees, they are investing in the company's future. This investment can payoff in several ways to include the following:

- *Increased productivity*: Employees who are current in the latest technology can facilitate productivity improvements through implementation of new technology and processes that can improve the way business is conducted and can reduce or eliminate potential bottlenecks.
- *Increase profits*: Employees familiar with the latest technologies and processes in their fields have the potential to stay ahead of their competitors and increase profitability.
- *Employee retention*: Employees who are offered the opportunity to continuously sharpen their knowledge and skills are more likely to remain with the company.
- *Employee feeling valued*: When leaders invest in employee training and knowledge retention, he or she feels valued and is comforted that management values him or her as an important contributor to the success of the company.

Continuing education is also an important way for professionals to keep abreast of their fields, so that they can enhance their knowledge and increase their marketability.

7.8 Knowledge Management

Today, many companies are experiencing a significant fluctuation in staffing due to a large number of workers retiring, moving on to work for other companies to take on other opportunities, or resulting from layoffs and downsizing. This fluctuation in staffing is creating an issue when it comes to retaining and managing the knowledge required for a business to retain its competitive edge. Knowledge management is being defined as the coordinated process of effectively capturing, evaluating, sharing, using, and distributing knowledge throughout a company.

Many companies are resorting to creative strategies to help them retain knowledge when they lose employees. Knowledge retention strategies can include the following:

- Focusing on retaining key employees
- Sharing information through social networking
- Developing and maintaining a current database of the key process and procedures
- Proceduralizing processes and system operations and maintenance

- Cross training staff to perform more than one job function
- Developing and using knowledge maps

Too often companies allow a worker to walk out of the door taking critical knowledge that has not been cataloged. In many cases, the knowledge that is going out of the door was gained through extensive training funded by the departing organization. Whatever means or strategy is used to manage knowledge so that a business can retain its competitive edge, the strategy must be one that can be implemented and sustained for the long haul.

7.9 Learning Culture

A learning culture is defined as a group of workers in the same environment sharing the same values for improving themselves and the organization through constantly increasing their skills and are continuously seeking performance improvement through gaining new knowledge and applying that knowledge to achieve the mission of the organization. The key attributes of a learning culture are pictured in Figure 7.4. These attributes are embodied in the attitudes, values, and the practices of the members of an organization working together to accomplish a common goal.

In a learning culture, organization members are never happy with the status quo. They are always seeking ways to make organization life better, improve the quality of its products or services, and continue to deliver value

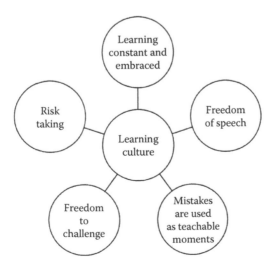

FIGURE 7.4
Key elements of a learning culture.

to its customers. Workers in a learning culture tend to exhibit the following characteristics:

- Freely share information and knowledge
- Are very inquisitive about how work is performed
- Openly provide honest feedback without fear of retribution
- Ask thought-provoking questions

7.10 Assessing the Effectiveness and Quality of a Training Program

Many training programs fail to deliver the expected results for employees as well as the company. Therefore, it is necessary to have a process to evaluate whether the training designed to educate and train workers is effective and achieve the expected goals. A well-structured assessment process can identify training gaps and areas that are in need of improvement. The effectiveness assessment only provides information on the issues that are preventing or limiting the effectiveness of the training program. In order to determine the root cause of the issues, a root cause analysis may be helpful to flush out specific causes that must be addressed.

One of the most common methods used to measure training effectiveness was developed by Donald Kirkpatrick. This model has been used by many training organizations and continued to be used today by many practitioners. The Kirkpatrick model is a four-level model as depicted in Figure 7.5.

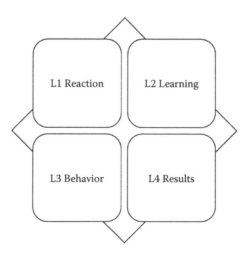

FIGURE 7.5
Kirkpatrick training assessment model.

The questions that are addressed by each level are as follows:

Level 1: To what degree did the trainee reacted favorably to the training finding the training to be useful?

Level 2: Were there improvements in knowledge and skill based on their participation in the training?

Level 3: Were the participants able to apply what was learned during the training in the work environment resulting in a change in behavior?

Level 4: Were there any benefits to the organization as a result of the training that can be measured such as productivity, efficiency in operations, or improved quality?

VIGNETTE 7.2

The following are two resources that can be used as an aid to evaluate training:

Kirkpatrick's 1998 book: *Evaluating Training Programs: The Four Levels*, 2nd Edition, Donald L. Kirkpatrick, Berrett-Koehler publishers, Inc.
Alston, F. and Emily, M., *Guide to Environment Safety & Health Management*, Chapter 8, CRC Press, Boca Raton, FL, 2016.

Another very important element in assessing training is to simply solicit feedback from the course participants. This is commonly obtained through feedback surveys or questionnaires immediately after training is completed.

7.11 Summary

Trained employees have the knowledge and skills needed to perform their day-to-day jobs. A good effective training program that addresses knowledge management and retention is an investment that cannot be overlooked in terms of value to a company's financial bottom line. An effective training program can be costly; however, it is necessary for business to have the resources needed to compete in a global economy. Training is not a one-time activity; therefore it must be a continual part of the corporate strategy.

It is the responsibility of management to ensure that a training program is designed with the appropriate resources to serve the needs of the company. Often, when funds become tight and management has to decide where budget cuts will take place, training is one of the first places where funding is reduced. Without continuing knowledge refreshment, employees are stuck in time and have very little opportunity to advance their knowledge in the work environment. In such cases, the employee may feel that the leadership team does not value him or her enough to continue to invest in his or her knowledge, career, and personal development. Because of its importance to the success of a company, a strategy for knowledge retention that includes effective training should be included in the overall corporate strategy. In addition, an effective training and development program is a great tool that has been associated with employee retention.

8

Environment Health and Safety in a Lean Environment

8.1 Introduction

Effective environment health and safety (EH&S) practices in conjunction with sound Lean practices can help organizations gain a competitive advantage in the business in which they compete. Typical areas that comprise EH&S include industrial hygiene, radiation safety, occupational health, conventional safety, and environmental protection. As corporate leaders are pressured to continuously improve their operations and reduce cost, they are looking to implement practices across the organization that will aid in this initiative. Lean techniques coupled with a sound EH&S program is a combination that will yield enhanced value across the board. Many leaders are placing their focus on process improvement in areas such as technology and manufacturing. A key component of the process of reducing waste and adding value is being overlooked that can pay dividends in many areas for a company. That area is the EH&S function of the company.

In today's corporate environment, the focus on waste elimination is necessary to continue to improve and garner funds that can be reinvested into the company. Leaders are beginning to recognize that the EH&S functions is an untapped component that can significantly contribute to the bottom line. Minimizing waste through an effective health and safety (H&S) program coupled with Lean thinking and the use of Lean techniques can help a company to retain current market position and potentially gain new business.

8.2 Components of Corporate Environment Health and Safety Program

EH&S programs focus on all aspects of human health in the workplace and protection of the environment. These programs are designed to address all of the physical, chemical, and biological factors that can be present in the workplace that can potentially cause harm to employees. The success of an EH&S program cannot and will not be successful without leadership and employee involvement. Management is expected to commit the resources needed such as staffing and funding to ensure that all workers have a safe environment to perform their jobs and facilitate a culture where employees feel empowered and safe to participate and communicate openly.

Visibility of management in leading the design, implementation, and continuous improvement, and support for the organization's safety and health activities is critical. Specifically, the highest level of management should establish and review the site's safety and health policy at some frequency, and should ensure that employees are familiar with the policy and the role they play in implementing the policy and management expectations. Key areas of an effective environment safety and health EH&S program include the following:

- Environment protection
- Worker safety and health (WSH)
- Medical care
- EH&S training
- Radiation protection—where applicable

Each of these areas will be discussed in brief in Sections 8.2.1 through 8.2.5. These rubrics are designed together to cover the needs of the company and employees to facilitate work in a safe and compliant environment.

8.2.1 Environmental Protection

All companies have a need for some form of a program to ensure compliance to the many environmental regulations that impact their business. These regulations are not inexpensive to comply with and at times may be somewhat complex in compliance structures. Regardless of the cost of compliance or the complexities in achieving compliance, these rules are the law. Therefore, regulatory compliance is not an option for companies. In developing a program to inform the needs of a company, it is essential that management allocates proper funding to support the initiative, secure the skilled staff needed to develop and implement the program, and ensure a process in place that will provide metrics to determine program effectiveness.

8.2.2 Worker Safety and Health

The field of WSH is very important in ensuring that workers are protected from exposures associated with chemical, physical, and biological stressors. The most important aspects of WSH are being able to recognize, evaluate, and control workplace hazards using knowledge in the areas of science, engineering, and math. The process depends heavily on having knowledgeable subject matter experts who have the skills to anticipate potential workplace issues and the ability to use the right tools to engineer out hazards to protect workers so that they can perform the role in an environment where they are not injured.

8.2.3 Environmental Health and Safety Training

Many federal, state, and local regulations require training of workers before placing them in a job-related role in an organization. In fact, some of these regulations are so specific that they provide an outline of exactly what the employee should be trained on as well as the frequency for the training. It is the responsibility of management to ensure that workers receive the training necessary to aid them in performing their job responsibilities efficiently and safely. An effective training program not only addresses the gap in training the new workers, it should also be effective in training current workers on how to perform new tasks, improve their leadership skills, or prepare them to take on different roles in the organization.

8.2.4 Health and Medical Services

Access to medical care for workers is a natural part of ensuring that their health is not being negatively impacted by their work environment. In an environment where workers are exposed to chemical, physical, or biological agents, a medical program is a must. Also, some of the regulations designed to protect workers from workplace hazards require employees to be medically monitored or to be placed in a medical surveillance program to ensure that the workplace is not impacting worker health. Whether a company has an internal health clinic or a contract with an external health provider, access to medical care is necessary for employees and is an integral part of a corporate program.

8.2.5 Radiation Safety Program

Not all business will need a full-blown radiation protection program. In fact, unless a company is competing in the nuclear or medical industry, it probably will not need an extensive radiation protection program. Many companies that have small radiation sources or conduct research with low level radioactive products may address safety of workers as a part of another program. For example, some organizations address the safety of nonionizing radiation exposures as a part of their WSH program.

8.3 Eliminating Waste through Compliance

Ensuring that all applicable regulations are adhered to is a great way to eliminate the introduction of waste that can cost the company a bundle. Regulatory noncompliance can be costly and can take enormous time to address when an out of compliance situation exists and is discovered by the regulator or an auditor. Utilizing the concept of Lean to bolster regulatory compliance is a concept that is not often considered by the management team.

The tools of Lean are yet another powerful way to address any potential regulatory issues that must be addressed at various stages of a project. For example, if a new technology is being deployed within a company to assist with production that has an impact on the air and waste streams, an air permit or a waste permit may be required based on state, local, or federal regulations. At the point of discovery, the necessary permit applications can be flushed out and submitted to avoid delay in project or process implementation avoiding unnecessary or cost associated with delays. On the other hand, if a permit is required to operate a process and if the permit is not obtained before project startup, potential actions can be imposed by the regulatory agency, which can result in potential costs such as fines, increased regulatory visits, or loss of employee and stakeholders' trust. The use of value stream mapping is an effective tool when used with other Lean tools to identify these issues early on. Value stream mapping is just one example of using Lean tools to improve implementation of compliance-based programs. There are many other tools that can be used to enhance the compliance posture of a company.

8.4 The Role of an Effective Environment Health and Safety Program

An effective EH&S program has an objective with a documented strategy that is designed to prevent accidents, occupational illness and injuries, and prevent negative impacts to the environment. An effective program should begin with a policy statement that has the full support of senior management. The policy should contain at a minimum:

- The overall objective of the program
- The commitment of management to protect the H&S of employees
- Management responsibility and accountability
- The EH&S philosophy of the company
- Employee accountability and responsibility

An EH&S program must include the elements required by EH&S regulations and standards to demonstrate compliance and achieve the intended results. Because the mission of companies differs, it is not reasonable to expect that one EH&S program can be completely transferred across companies without modifications. However, successful programs can be shared as a starting point. Some critical elements of a successful environment, safety, and health program are listed as follows:

- Management commitment and responsibility
- Individual responsibility
- Employee led (H&S) committee
- Job hazard analysis procedure
- EH&S rules and procedures
- Employee orientation and training
- Workplace inspections, audits, and assessments
- Reporting and investigating workplace accidents, injuries, and illnesses
- Emergency procedures
- Medical and first aid treatment
- EH&S communications and promotions
- Employee engagement

An effective H&S department is an asset to a company when seeking to reduce injuries; to lower the cost associated with injuries, noncompliances, and violations; and to enhance relationship with stakeholders and the community. Too often, an EH&S organization is not effective because of the limited staffing levels and professionals lacking the knowledge and experience necessary to be successful in the roles they are expected to perform. Often, workers are placed in positions as EH&S professionals and managers without any background, knowledge, or skills in the function. Companies that hire qualified EH&S professionals generally are able to implement a solid program that is valuable in ensuring that work is performed safely and helping reduce the cost associated with accidents, regulatory noncompliances, and injury and illness to workers. Combining Lean principles with the EH&S program is an effective way to strengthen both programs.

Lean principles are not about working faster; however, it is about working smarter. Planning work in a way that it can be performed safely is working smarter and the only way to work. An effective EH&S program is a proactive tool in reducing waste and unintended cost, and it compliments a company's desire to implement Lean thinking, practices, and tools successfully.

8.5 Environment Health and Safety Programs Value Proposition

In today's safety conscious workplace coupled with the many EH&S laws that have been passed to protect the workers and the environment, workplace safety is at the forefront of getting work done. However, some leaders are still viewing the EH&S component of the company program as a use of funds that can be contributed to other processes or products. On the other hand, some leaders have discovered that an effective EH&S program is worth the investment, and they continue to contribute to improvement in the program. The outcome of their investments has included the following at a minimum:

- Reduction of injuries and illness
- Reduction in the cost associated with injuries and illness
- Lower experience modification rates (EMR)
- Lower worker compensation costs
- Increase in confidence of workers, stakeholders, and customers

The leaders who are supportive of an effective EH&S program have been on the receiving end of dealing with and recovering from accidents that impacted workers and their families. They are too familiar with the issues that accompany accidents that go beyond dealing with the injuries and the cost associated with a potential stop work for the process that was involved in the accident.

8.6 The Reality of Workforce Perception

Another link between Lean and an effective EH&S program is the perception of the workforce. In order to effectively implement an EH&S, the involvement, buy-in, and perceptions of the workforce are important; the same is true for Lean implementation. The way employees perceive their management team and their ability to function as leader is a predictor of whether the employees will support an EH&S program or Lean process improvement initiatives. Perception influences our ability to motivate ourselves and the way we behave. The following are some ways that can be used to influence workers' perception:

- Ensure that workers are actively engaged in the project.
- Communicate openly and frequently.

- Seek input from workers.
- Map out the process and demonstrate value to the company.

Workforce perception is an indicator of performance and success of programs. Therefore, attention is needed in monitoring perceptions and ensuring that workers are kept informed of changes as they occur.

8.7 The Real Cost of Accidents

How do you determine the complete cost for an accident? This is nearly an impossible question to answer. And frankly, when an accident occurs, no one is generally considering the financial ramification at that moment in time. One of the most wasteful activities that can occur in the work environment is an accident. Leaders are not always cognizant of the high costs associated with an accident, because these costs are not easily captured or quantified. One can say that the cost associated with an accident is another form of hidden costs.

H&S initiatives have and still are considered by some leaders as an expense to the company that some managers would like to limit their support. Until leaders recognize that H&S is an integral part of their operations that can be leveraged to improve business processes, reduce waste, and aid in employee retention, the function will remain an untapped function that can be instrumental in waste reduction and continuous improvement. If the event or the accident involves an employee injury, some of the costs are typically handled through the workers' compensation process. Other costs that may not be considered include lost productivity and higher insurance premium paid by the company as a result of employee injuries. Some additional potential hidden costs of accidents include the following:

- Equipment downtime
- Potential lawsuits
- Equipment repairs resulting from damage that occurred during accidents
- Investigation time
- Overtime to address the issue or perform the work of an injured worker

The reality is that it is impossible to completely quantify the total cost of accidents to an accuracy level of 100%. Therefore, all attempts to avoid it are the desired path because a company can lose more than it will ever be able to account for.

8.8 Applying Lean Principles and Application to an EH&S Program

Lean implementation first began as a means to improve business outcome in the manufacturing sector. This concept and way of thinking have been driving business processes in the right direction for many companies successfully. Lean has far-reaching applicability to many other areas that are crucial to business processes and, if used, can be a valuable way of improving outcomes. One such area is in the development, implementation, and monitoring of a corporate EH&S program. Many times, these programs are seen as stand-alone, and at times some would say it is cost prohibitive and stifles productivity. The reality is that these programs are a must and are sanctioned by regulations promulgated by federal, state, and local regulations. Lean tools can be, and have been, helpful in evaluating and maintaining an effective Worker Safety and Health (WSH) program. An effective WSH program is focused on recognition, evaluation, and control of workplace hazards. Lean thinking and tools can be helpful during each of the stages. However, it is most useful during the recognition and evaluation phases of safety and health. Some of the Lean tools that are helpful during these phases are listed in Figure 8.1.

When developing the corporate EH&S program in an organization where Lean thinking is engrained, program implementation can become easier (Figure 8.2; Table 8.1). Lean thinking organization thinks of ways to integrate Lean practices within all aspects of the organization to include

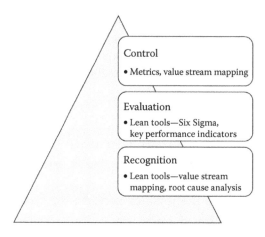

FIGURE 8.1
Tenets of worker safety and health.

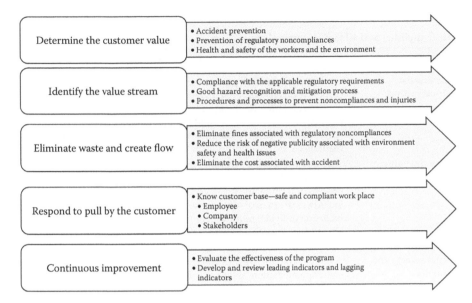

FIGURE 8.2
Overriding Lean principles and application to an EH&S program.

the way the organization functions. In fact, Lean thinking organizations try to stay away from developing stand-alone programs and systems. These systems and processes stand a greater chance of failure when implemented in silos.

8.9 The Synergy between Lean and Health and Safety Programs

Where do Lean and EH&S meet? The principles and tenets of Lean are closely aligned with the tenets of H&S. The challenge is that the linking is not often thought of or linked in such a way that the two functions are complementary in implementation. There are articles and books written linking Lean and a comprehensive and effective H&S program. Although limited, these bodies of literature are a demonstration that some leaders are beginning to recognize the link and the benefits of implementing the two programs in concert. Two examples of such literature are listed in Vignette 8.1.

TABLE 8.1

Lean Principles and EH&S Program Integration

Principles of Lean	Description
Determine the customer value	A well-designed EH&S program can and does add value to a corporate business strategy. Managers desire a program that is cost effective and provide the appropriate level of support to ensure protection of the workforce and the environment. Some customers may need assistance in defining the value and identifying what regulations are not applicable to their business if any.
Identify the value stream and then map	When a new process or technology is introduced and mapped to determine activities involved in creating and implementing the process or technology, it is a good business practice to consider and include in the map of all the associated EH&S requirements. Inclusion of EH&S requirement can save time and money down the line. For example, if a permit is needed it may take weeks, months to years to get it through the regulatory process.
Eliminate waste and create flow	An EH&S program can introduce waste through program development and implementation. When developing an EH&S program, ensure that the applicable regulations are accounted for and that the program is developed with the needs of the organization in mind. Avoid copying another organization program that has different business objectives. When implementing the program, ensure that the appropriate staffing is in place with the level of knowledge to aid implementation as not all regulations are clear, and some interpretation may be warranted.
Respond to pull by customer	Understand the applicable regulatory requirement and develop a process to facilitate compliance and avoid noncompliances that can result in penalties and fines.
Pursue perfection to achieve continuous improvement	Continue to review and evaluate the EH&S program and performance, and make adjustments as needed. It is important to keep abreast of regulatory changes as it will impact program performance and effectiveness.

VIGNETTE 8.1

One resource that discuss the application of Lean in developing and implementing a safety program using Lean concepts is listed as follows:

Lean Safety, Transforming your Safety Culture with Lean Management, Robert B. Hafey, CRC Press, Taylor & Francis Group, 2010, discusses the link between Lean and safety, and success of Lean is highly dependent of safety.

When reviewing the aforementioned books and other references that show a definitive link to the synergy and benefits of taking a Lean approach to H&S,

you will frequently hear the term *Lean safety*. Lean safety is simply defined as a systematic approach used for the express purpose of identifying and controlling the waste in processes and activities, which could cause or contribute to workplace accidents or illnesses.

8.10 Lean Tools Usage

There is a wide array of Lean tools that are used to improve H&S in the workplace as well as improve implementation of Lean in an organization. One important Lean tool that is used to organize the work area to improve workplace functions and deliver exceptional results in the area of safety, quality, efficiency, productivity, and cost reduction is the use of the 5s principle. The tools are designed to improve workplace efficiency by focusing on organization, standardization, and cleanliness. The 5s and its link with Lean are shown in Figure 8.1. Some organizational benefits of using the 5s include the following:

- *Improved workplace safety and health*: Workplace hazards can become visible by asking thought-provoking questions that can be dealt with prior to an accident or injury occurring. Some examples include removal of tripping hazards resulting from housekeeping, improper storage of hazardous materials, and the list goes on.
- *Improved productivity*: The 5s can facilitate the elimination of waste, thereby providing the opportunity to increase productivity and efficiency.
- *Encourage employee engagement and boost morale*: The 5s encourages participation of employees by providing inputs and suggestions to identify causes of accidents and to improve process and workplace safety.

The 5s is an effective Lean tool that can be used to improve H&S in an organization. The proper use of the tool can improve productivity and can create a safe and hazard-free workplace. The five (5) whys is another popular Lean tool that is used by EH&S professionals frequently in the root cause evaluation process to determine the cause of accidents, injuries, and other EH&S-related issues and to identify, control, and eliminate workplace hazards. There are many Lean tools that can be just as effective in application (Figure 8.3).

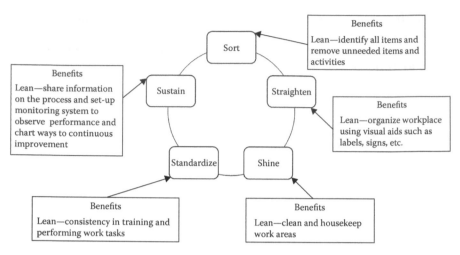

FIGURE 8.3
H&S and Lean synergy—5s.

8.11 Summary

Utilizing Lean thinking and tools in the development and implementation of an EH&S program is an innovative concept that has not been fully realized. Corporate leaders who are actively engaged in ensuring that the EH&S program is effective in recognizing that the cost of an effective program is a necessary expense. On the other hand, some leaders are skeptical of the value that the EH&S program brings to the company's portfolio and will contribute as little resources as possible to support the program.

The principles of Lean have been applied to many workplaces successfully in changing the way business is delivered to the customer. These changes have enabled leadership to focus on the products and services it delivers and to optimize the quality along with continuously improving the processes, practices, and procedures. These principles can also be used to optimize safety and health of the workplace. Combining Lean tools to enhance the effectiveness of an EH&S program is a great way to identify and address potential issues before they become real issues.

9

Implementation Pitfalls and Hidden Costs

9.1 Introduction

We have already emphasized that implementation of Lean is generally not an easy task and that it can often fail. However, success is feasible as long as the right steps are taken at the right time and step in the process. In many cases, some difficulty when encountered can stifle success if not addressed as early as discovered or anticipated. As managers are under tremendous pressure to continuously deliver quality and reduce cost, it is conceivable that some managers will commit to operating under the premise of Lean knowing that the organization may not be positioned for success. They will often seek projects that are considered to be *low hanging fruits* that will achieve some level of efficiency; however, the change achieved initially is not sustainable. Generally, there is nothing wrong with starting down the path of least resistance; however, even on this path, the correct culture must exist, the leadership team must be prepared, employees must be engaged, and there must be a commitment to the project and the process. Another action that can be helpful to leaders before they embark upon the Lean journey is to conduct a risk assessment. A comprehensive risk assessment can be helpful in minimizing and in some cases eliminating the potential for failure and reducing unintended costs later in the project.

Some managers often take credit for cost savings that are not entirely real or accurate. This can be in part due to the difficulty in quantifying some costs, and managers may be under pressure to show cost improvements. In this chapter, we are going to discuss many of the hidden costs that are not included in the expenditures and generally not reported when implementing Lean and other improvement initiatives. In addition, a discussion on ways to reduce these hidden costs and pitfalls will be highlighted. Although Lean can be successful for the long term, when quantifying the cost associated with the process, all costs are not always considered during the initial stages.

9.2 What Are Pitfalls and Hidden Costs?

The pitfalls and hidden costs of Lean are not always of a financial nature and not always quantifiable in some cases in dollars and cents. These pitfalls can exist because of actions such as poor planning, the organization is not ready for change, or the leadership team and employees are not supportive or engaged.

Some of the pitfalls that can lead to hidden costs of Lean implementation include the following:

- The cost of discovery
- Start-up cost (additional resources such as equipment and people)
- Loss of personnel
- Retraining of staff to the new process
- Displacement of staff because of the gained efficiencies
- Communication degradation
- Loss of trust
- Fear of job loss
- Revision to new policies and procedures
- Development and implementation of new policies and procedures
- Incompetent leadership

These types of hidden costs are not often accounted for or included in the cost of implementing Lean process improvement initiatives. Therefore, the benefits achieved versus the benefits communicated can be significantly distorted and not accurately quantified and accounted for. We will discuss in some detail in this chapter many of the pitfalls and hidden cost that may be encountered while implementing process improvement initiatives.

9.3 A Project Management Approach to Lean

One way to help reduce the additional cost and pitfalls associated with Lean implementation is to use a project management approach. Recognizing that not all projects are large enough to use a full-blown project approach, modification to the approach can however be made based on the project. Project management is an organized process used to oversee projects to ensure that the goals of the project are met while keeping in line budget, quality, and timeline for completion. The use of project management has a number of benefits that include the following:

- Providing a clear definition of why the project is necessary
- Preparing the business cases for the investment
- Developing and managing the budget
- Documenting and bounding the scope to help avoid scope creep
- Increasing the probability of completing the project and achieving the desired results
- Estimating and facilitating efficient use of resources
- Monitoring project performance to maintain the schedule and the project plan
- Monitoring expenditures
- Estimating project time line and management plan
- Monitoring and controlling risk
- Ensuring project closure in a systemic and controlled manner

Successful projects require careful upfront planning, and taking a project management approach facilitates an environment for effective planning and implementation.

9.4 Organizational Alignment

There are many ways to reduce the potential pitfalls of Lean failures. One such way is found in the organizational structure. The way an organization is structured sets the stage for

- The functionality
- The way the work is performed
- The communication channels
- The way leaders and workers interact

Therefore, it is advisable that when designing the structure of an organization, the right structure must be selected and put into place if Lean thinking and implementation are the goals for functionality of the organization. An example of an organization that can facilitate Lean is described in Vignette 9.1.

VIGNETTE 9.1

In an organization aligned with projects in mind, the structure takes advantage of using elements of a functional organization as well as a traditional matrix organization. This alignment is chosen primarily as a means to minimize duplication of resources and effort as well as to provide diversification of work experience for employees who support different projects.

Organization alignment has the potential to help chart the course of being able to implement Lean successfully with longevity. An organization aligned to support work with a project management design should make it easier for staff to adopt a flexible mind-set and be able to embrace change.

In a projectized organizational structure (Figure 9.1), the organizations are arranged based on activities into programs or functions and implemented through projects. In this structure, the project manager is completely in charge of the project and everyone in the team works under his or her direction, although he or she has a functional home organization. Important benefits of a projectized organization are that they are very adaptive to change, and members are able to learn from their mistakes as well as the experiences and mistakes of others across the company, because the workers who are used to staff the projects also most likely have had the opportunity to work on other projects across the company.

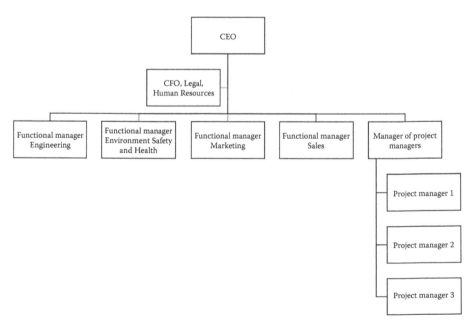

FIGURE 9.1
Projectized organizational structure.

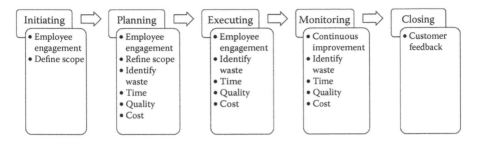

FIGURE 9.2
Project management approach to Lean.

One can expect in a projectized structure that each team is staffed with subject matter experts from the functional organization and assigned to a project manager to work on a specific project. Once the project is complete, the subject matter expert may move on to another project managed by a different project manager or remain with the current project manager to complete another project. The SMEs are assigned based on the resource needs of the project. In some cases, an Subject Matter Expert (SME) may split time between more than one project. In such cases, a schedule must be agreed upon between the project managers to ensure that their projects remain on schedule, and the appropriate level of quality support is delivered by the SME for each project. This arrangement is not optimal but can work with planning and coordination. Benefits of the project management approach to Lean are further described in Vignette 9.2 and Figure 9.2.

VIGNETTE 9.2

Lean implementation using the project management approach helps to control and monitor project elements, such as budgets (reduces the potential of overrun and hidden costs), resources, and time to completion and to ensure success of implementation.

9.5 Why Lean Fails?

There are various reasons why many attempts to implement Lean fail. Most of these reasons are as follows:

- Lack of commitment from the leadership team
- Lack of employee engagement
- Culture of the organization not appropriate
- Lack of infrastructure to support the new process
- Failure to address the people aspects

- Selection of a technology that is ineffective for the business
- Failure to implement project management approach and tools
- The leadership teams are not Lean thinkers
- The strategy does not compliment the project vision
- Trying to solve the wrong problem
- Tools used did not identify the problem accurately
- No customer focus

However, it is still feasible for a Lean project to succeed if not all of the attributes are visible during implementation. Although many would contend that the lack of commitment from the leadership team is the single attribute in itself that can lead to failure. One hidden cost that may not be considered is all of the cost associated with implementing a process that quickly or eventually fails. The cost associated with these activities is definitely hard to quantify or to some extent unquantifiable. Therefore, these costs are generally not captured or reported.

9.5.1 Performance Metrics

Measuring performance is a key to evaluating how well a company is performing and meeting the needs of its customers. Performance metrics, if designed appropriately, give hard data that answer performance questions. Good performance metric will yield results that clearly measure defined areas and attributes within an identified range that allows for continuous improvement. Attributes of good performance metrics are shown in Figure 9.3.

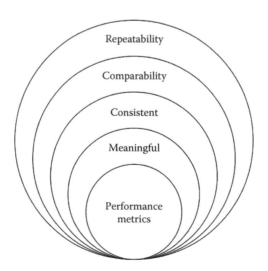

FIGURE 9.3
Performance metrics attributes.

Metrics can be used to

- Determine if customer satisfaction is being achieved
- Determine if the group or company meets commitments
- Determine if quality is maintained
- Highlight continuous improvement opportunities
- Uncover areas for cost reduction
- Improvement in cycle time
- Show productivity improvement areas

Performance metrics is a valuable tool to chart progress of a project that is in the implementation stage and more importantly a great tool to use to determine progress on a continual basis. Improperly designed metrics can give the indication that all is well when in fact systems are not functioning as designed. The issue with developing metrics is that it must be designed to capture the right data that will truly highlight performance and areas for continuous improvement.

9.5.2 Retention of Staff in a Lean Environment

Implementation of Lean depending on the project and the way it is handled by management and the workers can create some level of uncertainty for some of the existing staff. Lean is often viewed by many as a cost-cutting measure that is used by management to reduce cost that often leads to reduction in staff. Employees often think of job lost when they hear or think of process streamlining or reducing expenditures. This is not unusual because of all of the publicity a company receives when attempting to create additional value for its customers or stakeholders that results in a job loss or reducing worker benefits. These activities usually generate fear and distrust for the leadership team and the company once employees have an inclination that changes are on the horizon that may impact their current employment status with the company. Some employees may immediately begin seeking employment with another company. This can present a problem for the organization especially if the loss of critical resources is at jeopardy. To avoid or minimize the potential of losing critical staff when making changes and implementing Lean, consider the following:

1. Ensure that the workers are a part of the project from conception to implementation.
2. Communicate openly and honestly throughout the project.
3. Discuss openly any assignment changes that are anticipated.
4. Communicate and reassure workers that they will have a continued role in the organization.

5. Offer assistance to all workers who may be displaced as a result of operational changes.

6. Communicate the reason behind the revised process and the vision of the company.

The cost associated with losing staff or laying off staff, if this is the case, is not generally included in the cost of implementing Lean. The cost of replacing staff can be costly considering activities such as

- The time and resources needed to conduct talent search
- The cost associated with interviews
- Relocation cost for the successful candidate
- Sign on bonuses if applicable

And the list can go on and on.

When employees leave a company, in many cases, it tends to create stress for the workers who are left to perform not only the work that they have been hired to perform, but they take on additional work to cover for the employee who is leaving. This additional work can overwhelm the staff, and it can create additional cost that may not have been considered especially if workers are being paid to work overtime to accomplish the additional work, or if workers are taking additional time away from work because of issues such as fatigue or stress-related illnesses resulting from job-related activities.

9.6 Minimizing Risk through Lean

The purpose of the risk assessment process is to remove hazards and obstacles or reduce the level of its risk by instituting control measures. In doing so, a productive and safer workplace results. When implementing Lean, a comprehensive risk assessment can be essential in discovering and evaluating the potential risk that may hinder the project or process success.

Having the best people to execute the best thought-out plan does not alone guarantee success. There are a host of factors that can play a role in determining the outcome regarding whether a project has been successful or not. These factors are known as risks. Simply put, the definition of a risk is an event or occurrence that may negatively impact the project. Once known and understood, risks can be mitigated and often prevented. Addressing risk requires a good understanding of the project, product, or process; a good understanding of the risks; and advance planning. The components of risks can be multifaceted with input from many sources. Each component must be balanced in order to minimize the total impact of risk to a project, process, individuals, or the company. Some risk components that should be considered are listed in Figure 9.4 and briefly defined in Table 9.1.

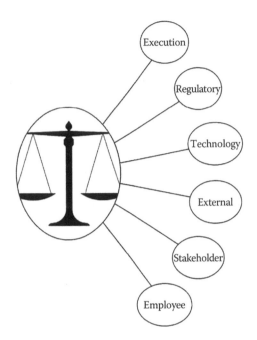

FIGURE 9.4
Component categories of a comprehensive risk profile.

TABLE 9.1

Risk Component Description

Risk Component	Description
Stakeholder	Individuals with any type of vested interest in the performance of the company, task, project, or product. Some examples include regulators (local, state, and federal), customers, suppliers, and public.
Employees	Company employees. Including full- and part-time employees.
External	At times, the execution of a project requires help and support from outside entities such as vendors, suppliers, or contractors. The dependence on these entities often poses risk to the execution of the project. To further complicate the issue, there may be little or no credible avenues that can be used to predict issues arising from these external sources.
Regulatory	An organization can be subjected to several kinds of regulations from the local and state government where it operates. In addition, many federal regulations may apply.
Execution	A project also faces risk generated from not receiving continued support from the organization and its leaders. Other risks may include: The organization may decide to redirect resources to other functions; project scope inadequate causing unanticipated costs; project may be abandoned because it becomes a financial burden for the company.
Technology	Often, project implementation involves the ability to acquire and implement new technologies. Technology risk can arise when the company does not have the ability to obtain the technology and the resources needed to fully implement the technology.

TABLE 9.2

Soft Risk

Potential Risks	Potential Impacts
Leadership team	Lack of support or the ability to manage the project to completion.
Employee involvement	Employees are not involved or supportive of the project.
Stifled communication	Lack of communication can create anxiety for the project staff as well as the workers who will be impacted by the project. This anxiety can lead to a lack of focus and support of employees and even contributing to some opting to leave the company.
Corporate culture	In order for a project to be successful, the culture must be supportive and appropriate for implementation and maintenance of the project.
Project or technology not well evaluated and implementation planned	Market analysis and proposal activities are not completely accurate, and project marketability and feasibility does not lend itself to implementation and profitability.
Human resource (HR) practices and policies	HR policies and practices may not completely support project as currently written. Polices may need to be revised.
Failure to anticipate and recognize changes in market trends	Market trends can indicate that the project may be obsolete or changes to the project or technology may be necessary. Ignoring market trend can lead to reduce marketability of product or technology.
Knowledge development and retention	Unable to develop and retain employees needed to contribute to the project success.
Budget exceedance	Budget exceedance can result from broad unbounded scope, scope creep, schedule and planning issues, as well as other activities that contribute to budget exceedance.
Leadership team relationship with workers	The relationship between leaders and workers can determine whether a leader is able to tap into the creative thinking of workers that is needed to improve business outcomes.

Some risks that are not always included in the traditional risk assessment are listed in Table 9.2. These risks are not always easy to account for, since they are based on the actions and reactions of people. These risks are being referred to as soft risks.

9.7 The Impact of Trust on Costs and the Pitfalls of Lean Implementation

We have touched upon at varying levels throughout the book that trust is an important attribute in developing culture and implementing Lean in organizations. Trust is at the core of everything we do when it involves following the direction of another person. It is an important element in personal and work relationship building. In both cases, trust can, and often it does, determine success or failure. In this section, a more detailed discussion of trust and the impact it can have on decreasing the feasibility of a project's success as well as the hidden cost that may not be so obvious is provided.

VIGNETTE 9.3

A detailed discussion on the impact of trust on organization and its performance as well as avenues and instruments that can be used to improve worker's trust is discussed in the following book:

Alston, F., *Culture and Trust in Technology-Driven Organizations*, CRC Press, Taylor & Francis Group, Boca Raton, FL, 2014.

Because trust in people is emotion based, it generally cannot be quantified in dollars and cents although the financial impacts can be real. Workers are willing to contribute to the organization's mission and goals when they trust their leaders. In a work environment where trust is limited or does not exist, employees are not willing to go the extra mile to ensure success is achieved when implementing new processes, practices, or technology. We can view trust as an investment that is sure to lead to improvement in business outcomes.

9.8 The Impact of Culture

The impact of culture on Lean implementation is mostly in the category of those costs that are difficult to quantify. In many cases, these costs cannot be quantified because the effects are based on environmental conditions that impact the way people act and react to changes. Although difficult to recognize and quantify, culture plays an important role in whether process changes are a success or a failure. The mistake that many managers make is that they do not consider culture as they are planning organizational and process changes. Culture can be one of the most significant attributes as to why projects fail. Therefore, before implementing organizational changes, consideration and an evaluation of the culture are important and can provide information on whether the timing is right to implement Lean or any other organizational changes. A culture survey questionnaire has been included in the resource packet in Chapter 11 that can be used to aid in diagnosing the status of the culture operating in your organization. Equipped with this knowledge can save the company significant hurdles that can equate to financial loss. Additional information on the impact of culture on Lean implementation is discussed in Chapter 3.

9.8.1 The Role of Subcultures

Most organizations have several pockets of subcultures operating within the organization. This is especially true for large organizations. There could be as many subcultures as there are department or work groups. Subcultures are a segment of the overall culture that exhibits the existence of different customs, norms, values, practices, and beliefs of what is found within the organization as a whole. This difference can result from the differences in department goals, management structure, job requirements, or other factors.

Organization subcultures can have a significant impact on the overall culture of an organization. In fact, depending on the strength of the subcultures, it can change the culture of the organization and impact successful implementation of programs and process to include Lean. On the other hand, if a process is slated for implementation in an area or group with a subculture that possesses the appropriate attributes, success of the process in that group may be enhanced while failure may be seen in other subcultures within the same organization. The point is that subcultures can be at the helm of Lean failure or success. The cost associated with subculture impact on Lean failure is real, yet not easily quantifiable; therefore it is not generally included in any cost analysis and estimates.

Strong cultures can facilitate and render a level of stability in an organization, and this type of stability can have an impact on productivity. In order to successfully manage and implement projects within an organization, leaders must be able to create and maintain effective subcultures as well as the culture of the organization.

9.9 Lean Procurement

Implementing a Lean procurement practice and process has the potential to dramatically change the way the business is conducted in a variety of companies. Often, waste is generated at various stages of the procurement process when implementing traditional procurement systems. Lean procurement refers primarily to the implementation of a standardized process to procure materials and supplies using a process such as just-in-time inventory practices.

Successful Lean procurement means that quality is not jeopardized with no increase in time and cost. Caution is necessary when implementing a Lean procurement process to not assume that unintended costs are not being generated, and that all cost elements are known. Potential hidden costs can often be seen in the implementation of a just-in-time process that is not functioning properly. An improperly functioning procurement and supply chain process can add risk that sometimes translates into unantici-pated costs when

- Suppliers are not able to adjust quickly to an increase in the supply demands.
- A supplier experiences distribution problems.
- The company is unable to meet the demand of delivering on massive nonroutine orders because it stocks fewer supplies.
- Supplies are not ordered and received as needed resulting in workers waiting around for supplies in order to be productive.
- A supplier does not deliver products on time and in the correct amounts impacting the production process.
- The investment cost of linking the company to its supplier in order to cordinate and facilitate on-time delivery of materials.

9.10 Hidden Costs Are Multifaceted

When speaking of the hidden cost that may be involved in implementing Lean or any other process improvement initiative, reference is not only being made to the financial aspects. Hidden costs can involve anything from the impact of Lean on workers, the leadership team, and monetary investments. Additional hidden cost can include the following:

- *Productivity impacts*: The work that was being performed by the worker who left the company for the fear of job loss has to be reas-signed to someone else in the company, or a new person must be hired to take on the scope. Most likely, productivity will decrease due to this change in work assignment for an internal worker and more so for a newly hired employee.
- *Lost knowledge*: The knowledge that the person who is leaving the company goes along with him or her. The training provided to the departing worker is no longer considered a benefit; it becomes a loss to the organization.
- *Recruitment and interviewing costs*: Although your company may have dedicated workers to perform the task of recruitment for the organization, there are still additional costs that may not be included in performing that function when seeking to replace workers. Costs such as advertisement and transporting perspec-tive candidate to an interview location (travel and relocation costs).

- *Training costs*: Training cost may include additional classroom training and on-the-job training. It is easier to quantify classroom training; however, it is extremely difficult to quantify on-the-job training expenses.

9.11 Capturing the Actual Cost of Lean

There are some actions that should be considered when implementing Lean to provide an indication that will help ensure that the project is financially worthy and that real cost savings will be realized and reported. Some actions to consider are listed as follows:

1. Link and monitor the Lean strategy and the improvements being sought. Ensure that the project will yield the intended financial benefits and will not create unintended cost.
2. Ensure that an accountant is embedded in the team and in attendance at planning meetings. The accountant can be instrumental in helping the team translate improvements to operations to financial impacts.
3. Develop and implement financial guidelines early on in the project.
4. Time is money. Ensure that the project is well defined with defined timelines.

Lean can become a lost opportunity for many organizations because they are unable to translate the process improvement into sustainable improvements and financial gains. When calculating the bottom line, they see financial loss resulting from a variety of unanticipated activities that were not considered in the upfront-planning process. In these cases, management may begin to question the value of proceeding with Lean if it is unable to realize financial or productivity gains. One practice to consider when embarking upon the Lean journey is to develop a cost mapping along with the initial map that will be used to show the *to be process*. This will serve as a point to begin quantifying the true cost of implementing Lean as well as the unintended cost that may be generated at the various stages of implementation. With the cost analysis plan in plain view, some of the unanticipated cost can be avoided or accurately captured.

9.12 Summary

As we have discussed in this chapter, there are a lot of ways that cost can accumulate when implementing Lean. We have also established that not all of the cost associated with implementation is always accounted for and quantified. Although hidden costs are difficult to quantify, this does not mean that an honest attempt to capture these costs should not occur. It is important to be able to capture the complete cost of a project in order to determine whether the project is cost prohibitive or if moving forward makes sense. The hidden costs associated with Lean implementation can leave a bad taste in the mouth of management when the project is implemented, especially if the project is not delivering the anticipated gains in finance, quality, and efficiency.

10

Lean Implementation Case Study

10.1 Introduction

In this chapter, a detailed discussion of a Lean project will be introduced along with the strategy, unanticipated cost discovered, and pitfalls that had to be addressed and overcome. Also, the resolution of issues encountered throughout the project will be presented. As realized by many, Lean implementation is not easy. However, it is feasible to successfully implement Lean strategies as long as issues are anticipated, evaluated, and addressed early on during strategy and implementation. This case study is offered to demonstrate processes that can be used to effectively implement Lean in an organization. As we have mentioned many times before, considerations must be given to the people aspects prior to implementing Lean concepts. You will note that the project represented by the case study was successful primarily because of the following:

- Extensive employee engagement
  - Recommendations that solved real solutions and generated efficiencies were offered up by the workers.
- Workers brought into and supported the strategy offered up by management using their input.
- The trust workers had in their management.
- Lean thinking leadership team.
- A well thought-out strategy.
- The appropriate checks and balance in place during planning and implementation.
- Anticipating and addressing issues early.
- The culture was optimal for implementation.
- An extensive communication campaign.

10.2 The Project Description

The project involves implementing a process that is used to track chemicals used by a large company from cradle to grave. The practices used previously were more of a hands-on process that subjected the workers to hazards that could arise from handling a large variety of chemicals in varying sizes, weights, and container configurations. In addition, the process created ergonomic-related challenges for the worker as he or she began to have visible health impacts.

The task of inventorying chemicals for the entire facility is time consuming, and it required the workers to place themselves in stressful positions that led to various ergonomic-related injuries. Some of the workers began taking time away from work due to the ergonomic-related injuries. This spark recognition of the need to seek out a different method could be used to accomplish the task. In order to address the issue, management solicited the assistance of the workers. As you would imagine, the workers were eager to help, and they had already formulated some thoughts on what they thought would be a viable solution. Thus, the search began to identify a new process for inventorying chemicals from cradle to grave, create efficiency in completing the task, removal of potential hazards for workers completing the task, and limit disruption for customers while workers are in their work environment completing the inventory of their chemical products.

10.3 Identification and Quantification of the Problem

When deciding to implement any type of process improvement, it is necessary to first quantify the problem that is being addressed. There are various ways to identify the *as is* condition and determine the *to be* condition. The *as is* condition simply is the current way in which the process is configured, and operates and the *to be* condition is the way in which the process is expected to operate to improve efficiency, safety, quality, and value for the customers. The avenues or processes that should be used to identify the issues are dependent upon the problem or process needing to be improved. The problem identification methods that were used during this project are listed in Table 10.1. There are other problem-solving methods that can be effective in identifying and resolving, which are not listed in Table 10.1.

It may not always be necessary to use multiple problem-solving methods to identify issues and solutions. However, in this case, several methods were used because there was a need to ensure that all issues and solutions were identified because the safety and health of workers were at stake. It is recommended that in some cases, the use of at least two methods may be a

TABLE 10.1

Problem Identification Methods

Method	Usage
Process mapping	A process map is a structural analysis of the flow of a process that shows all process-related activities. The ultimate goal of the map is to provide an overview of the relevant actions and procedures used to complete activities or tasks, so that organizations and individuals participating in a process are able to understand roles, procedures, and practices in the overall structure.
Brainstorming	Brainstorming combines an informal approach to problem solving using a team of people engaging in informal discussions. It encourages people to come up with thoughts and ideas that can provide solutions to various types of problems. It is most effective when the team consists of workers at varying levels within the organization who have hands-on knowledge of the task or activity as it relates to the process.
Employee interviews	Employee-based discussion used to understand the actual way work is being performed, and it provides input on best management practices and potential solutions from the view of the employee.
Root cause analysis	An effective problem-solving questioning technique that seeks to answer the question of why the issue occurred. The causes once discovered usually will fall into one of the three categories: physical, human, or organizational.
Task hazard analysis	A task hazard analysis is a technique that focuses on evaluating each task associated with a job or project as a way to identify all potential hazards. The analysis itself focuses on the relationship between the worker, the task, all associated tools that will be used in performance of work, and environment in which work will be performed. In order for a task hazard analysis to be effective, the workers performing the task must be included in the hazard analysis process.

conservative way to ensure that problems are identified, and the appropriate solutions are implemented. It also provides additional assurance and justifies the financial investment to others.

The case study issues were identified primarily through the use of brainstorming, process mapping, worker interviews, and the performance of an informal root cause analysis in concert with a task hazard analysis. The root cause analysis and the task hazard analysis were used together and were helpful in identifying and resolving the ergonomic-related issues that were encountered. Each of these methods will be discussed in further detail.

10.3.1 Brainstorming

Brainstorming is one of the most common practices used to explore solutions to issues using informal data gathering. Brainstorming with a group of people is a powerful technique and a good tool that is used when seeking ideas and discovering possible solutions to problems. Through brainstorming, new ideas are created, problems are solved, and team cohesion is facilitated and strengthened. Brainstorming places significant responsibilities on the facilitator to manage the process, team member involvement, and then to manage all follow-up actions. Effective brainstorming must be accompanied by selecting the right facilitator and team members; otherwise the process will not yield optimal results. When selecting team members, the model shown in Figure 10.1 can provide guidance on the selection criteria.

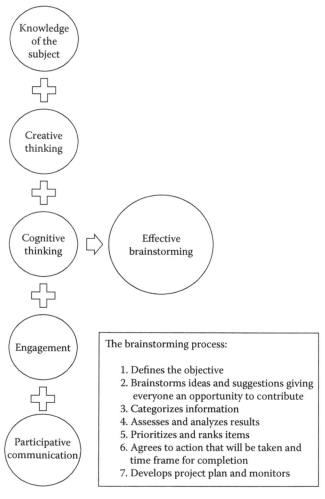

FIGURE 10.1
Brainstorming.

The two most common brainstorming exercises are as follows:

1. Each participant writes down his or her ideas on a sheet of paper. This approach may not be as effective as option 2.
2. Informal discussion among a group of people seeking solutions to a common problem and exchanging ideas.

The form of brainstorming used by the team for this project was informal discussions that are defined in item 2. This form of problem identification method was effective in helping the team in determining what issues they were attempting to solve. The questions that were at the helm of the discussions are listed in Figure 10.2. These questions served as an effective means to

- Reassure workers that management is concerned about their safety and health.
- Spark conversations between worker and management.
- Engage workers in identifying solutions to the problem.
- Reinforce to workers that their opinions are valued and listened to by management.
- Reassure workers that process improvement will not lead to job loss.

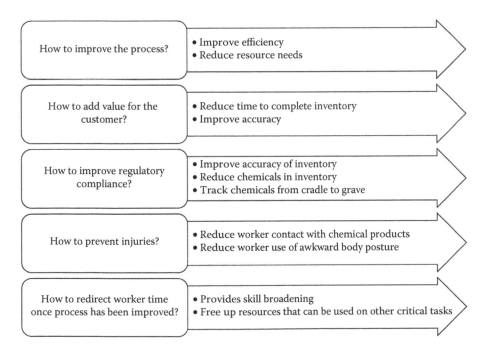

FIGURE 10.2
Brainstorming primary questions.

As a result of the workers being integrally involved in the process, many of the defined solutions, interim and long-term, were suggested, developed, and implemented by the workers.

10.3.2 Process Mapping

Process mapping is one of the simplest techniques used for identifying waste, streamlining work, and improving efficiency. Process mapping is also referred to as flowcharting, value stream mapping, or process charting. When constructed, the map shows the sequence of activities and events to complete a task and the outcome. Process mapping also has the ability to provide information on resource needs and cycle time. Benefits of process mapping include the following:

- Engaging employees and build consensus
- Uncovering of waste
- Facilitating process streamlining
- Highlighting areas where standardization can occur
- Helping participants to gain a thorough understanding of how the process functions

Process mapping is a good way to identify all of the steps that are performed to determine the *as is* condition. Once the *as is* (Figure 10.3) condition or process is known, it is easier to determine what the *to be* (Figure 10.4) condition or process should look like. Process mapping was used to determine

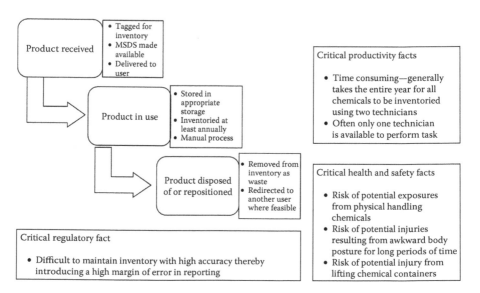

FIGURE 10.3
As is process.

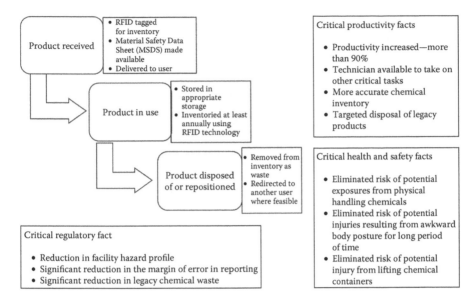

FIGURE 10.4
To be process.

the current process flow and provide information on areas that needed improvement to increase safety of workers and efficiency for the customers in completing and tracking their chemical inventories. One thing to avoid when developing a map is to avoid overcrowding. An overcrowded process map is difficult to read and buy into. To ensure the process is mapped sufficiently, ensure that an experience facilitator is used.

There were some process efficiencies made to the *to be* process that was not a part of the current process. However, the biggest change is in the way in which the inventory is being conducted, which minimized potential risks of injury and improved productivity.

10.3.3 Worker Interviews

Workers were interviewed to ascertain how they performed the task and to gain an insight into the difficulties they faced in completing the task. The workers willingly provided insights into task performance and ideas on what they thought could be done to improve the process used as well as what can be done to eliminate the ergonomic- and chemical-handling hazards. Workers answered questions such as

1. Describe the process used in completing the task.
2. Do you have all of the tools required to perform the task safely?
3. How often do you perform this task?
4. How long does it take to perform the task?

5. How often is it necessary to place yourself in an awkward position to complete the task?

6. How often is it necessary to lift containers weighing more than 25 pounds?

7. What are some of the issues encountered during task performance?

8. Are you experiencing health issues as a result of performing the task? If so, what type of health issues?

9. How can we improve the safety of the task performed?

10. How do we add value for the customer?

11. What recommendation can you provide on ways to improve performance of the task?

The answers to these questions formed the basis for many of the process improvement activities and changes embarked upon to improve safety, efficiency, and increase customer value.

10.3.4 Job Hazard Analysis

It has been demonstrated that workplace injuries can be prevented by evaluating workplace tasks and operations before beginning work. The best way to determine and establish proper work protocol and to ensure that the protocol is appropriate for safe performance of work is to conduct a job hazard analysis. A comprehensive job hazard analysis is likely to result in a safer work environment, fewer worker injuries, reduced workers' compensation costs, and increased worker productivity.

A hazard analysis can be accomplished by

1. Selecting the job or task to be analyzed.
2. Determining the scope.
3. Breaking down the job into sequential steps.
4. Identifying potential hazards.
5. Identifying preventive measures or eliminating hazards.

Questions that can be asked during the hazard analysis process include the following:

1. What is the scope of the task?
2. What can go wrong during the task?
3. How likely is it that a hazard will occur?
4. What steps or control can be put into place to prevent the issues that can create a risk to the workers?
5. Are there any environmental hazards that are not associated with the task that must be addressed?

The manager of the operation requested that the worker safety and health group performed a thorough analysis of the task and provided recommendations on how to eliminate the ergonomic-related risk to the workers. An important thing to note is that the workers were completely involved in the process of identifying risk and proposing improvements in completing the task. Employee engagement was high throughout the process and was invaluable in helping to identifying critical risk factors that led to the ergonomic stress and the increased risk of injuries. The job hazard analysis process identified solutions that could be implemented in the interim until a more permanent solution could be implemented.

10.3.5 Work Observation

Observing work is a great supplement to the job hazard analysis because it provided the reviewer a chance to see how the work is actually being performed by each worker. Work observation can provide an understanding of the hazards that are and can be encountered during the performance of the job. It can also provide an opportunity for the reviewer to recommend adjustment to the way the task is being performed, additional control measures, or ways to increase efficiencies. There are some fundamental actions to look for before and during work performance that are critical to the identification and determination of beneficial improvements. Actions to look for during the work observation process include the following:

1. Was a prejob briefing performed?
2. Was the scope discussed in detail to allow the workers to understand and reaffirm the limitation of the scope (bounding)?
3. Did the workers performing the task participate in the prejob briefing?
4. How did each worker perform the task?
5. Did the worker appear to have an understanding of how to perform the task?
6. Were safe work practices used by each worker?
7. Did the worker have the tools needed to accomplish work safely?
8. Were there any safety-related issues observed?

The work observation was performed by a qualified health and safety (H&S) professional and resulted in actions that help to reduce the potential of exposure for performing the task. The H&S professional agreed with management and the employees that a permanent solution that included a change in process was warranted. The recommended changes were made immediately.

10.4 Benchmarking and Technology Selection

Once the decision was made to seek other ways of performing a task or process, a good way to determine the processes that are available to address process improvement is to benchmark other organizations' performance of the same task. Benchmarking is a great way to look at alternate ways of doing things and learning of the challenges experienced by others who have used the respective process. Benchmarking was conducted, and it provided clarity that the use of RFID would provide a safe and efficient means of tracking chemicals from cradle to grave.

RFID stands for radio-frequency identification and refers to small electronic devices that consist of a small chip and an antenna that is capable of transmitting up to 2,000 bytes of data. RFID devices serve the same purpose as using traditional bar code; it provides a unique identifier for the object it is connected to. However, it was determined that RFID would work better than barcodes for this application because RFID devices do not need to be positioned in direct proximity of the scanner, and these devices will work up to 20 feet away from the scanner. The RFID technology was selected for the application, and the concept was presented to senior management for approval and buy-in.

10.5 Conducting the Pilot

Pilot studies also known as feasibility studies and proof of concept are conducted using a small-scale study of a project. These studies are used to provide critical information on what will occur during the full-scale project. Pilot studies provide conformation on attributes such as

- Feasibility of the process or procedure
- Time required to complete
- Cost
- Size of the effort
- Uncovered potential issues so that they can be addressed before full project implementation

The pilot study conducted before beginning the full study provided meaningful information that was used to develop the full-scale project before implementation. The information gathered during the pilot also provided the basis for the presentation to senior manager in seeking support for the project.

Some of the information that was obtained or validated included the following:

- The amount of financial support needed to complete the project
- Time required to complete the project along with the necessary resources needed
- Validated productivity and potential cost savings expected
- Demonstrated reduction in the risk of potential injury and illness
- Highlighted adjustment needed to the project strategy to ensure success

10.5.1 Implementation of Technology to Reduce Risk in Conducting Chemical Inventories: A Case Study

The pilot study provided critical information on attributes that needed to be changed to ensure success of the full-blown project. Based on this information, the project was revised and ready for implementation. Essentially, implementation was placed into three stages to ensure success. The stages were as follows:

1. Institutional support
2. Resource gathering and training
3. Equipment procurement
4. Process conversion

With the pilot completed, the benefits validated, the cost, resources, and time to complete the project known, it was now time to gain support from institutional leaders. Presentations were provided to leaders from various groups to incorporate and understand the voice of the customer, to introduce the concept, and to gain support for the project. Many customers overwhelmingly provided their support for the project. Once the prospective of the customer has been gained, a presentation was provided to the senior leadership team to gain support and funding. After institutional support was gained and funding was committed, the project was on schedule to begin.

The next step was to determine the resources needed to complete the project within a one-year time period. To ensure that the appropriate workers were assigned with the needed physical capabilities and skill sets, a job demands that a work sheet was developed and provided to the resource manager. The selection process took longer than expected; however, it was effective in ruling out the candidate who was unable to perform the task successfully. The selection process impacted the schedule, and the schedule was rebaselined taking into account the extension needed to select and train candidates.

Training for some workers took longer than expected. At this point, the realization of yet another potential extension on the project was realized. Once the workers had been trained, and procured equipment was in hand, it was time to begin conversion of the chemical inventory using RFID technology. The process began initially with dispatching two teams, and later additional teams were added to ensure the project would remain on schedule. The cost associated with the schedule extension and the extended time needed for training were not included in the original cost estimates.

10.6 Lean Thinking Leadership

Previously, we talked about the attributes and characteristics of a Lean-thinking leader. In this section, the focus will be on the leader for the group who was responsible for completing the chemical inventories in the case study. We often talk about whether or not good leaders are born or developed. In the case of this leader, he or she evolved over time into an exceptional leader with the right balance of the four prominent characteristics exhibited in the diagram shown in Figure 10.5 and discussed in Sections

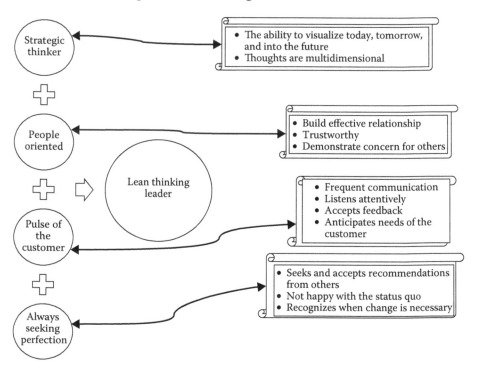

FIGURE 10.5
Lean thinking leader paradigm.

10.6.1 through 10.6.4. Let us explore how those characteristics played a role in the project from conceptualization to implementation.

10.6.1 Strategic Thinking Leadership

A leader who thinks strategically has the ability to visualize the future and influence others through the way he or she conducts business in an orderly manner. These leaders are not typically the type of leader that is known for *dropping the ball*. Therefore, the worker sees continuous team movement and improvement in the way business is conducted. When workers see constant movement toward improvement in business, they tend to want to become a part of the movement. They themselves become strategic thinkers because they have a good role model in their leader. The one caution that a strategic leader must constantly be aware of is that he or she can overwhelm others who are not strategic thinkers with the way he or she thinks through and chart the course for success. The leader for the project strategically planned the project from conception to implementation, actively engaged the workers who performed the task in identifying the problem and solutions, and sought early on and obtained support from management, workers, and customers.

10.6.2 People Oriented and Relationship Builder

Building relationship with others remains an important attribute of a good leader. A leader must have the ability to relate to workers as well as clients. The relationship leader has the ability to bring out the best in people because he or she is viewed as being trustworthy and is generally trusted by workers as well as customers. They are attuned to the expectations of others and the way they interact with others. People-oriented leaders are skilled in inspiring people to place their confidence in them and support their vision and projects. Other characteristics of this leadership style include the following:

- The tendency to energize employees because they make people feel appreciated for the work they do.
- The focus on employee relationships makes employees feel that they are valuable to the mission of the organization.
- Workers feel safe to be engaged in business activities and provide feedback willingly.
- People feel that they are a part of a company's success.

The manager for the project team was skilled in building and maintaining relationships with workers and customers. The relationship and demonstrated concern shown for workers was key to the level of worker involvement and buy-in for the project.

10.6.3 Pulse of the Customer

Developing and maintaining customer relationship is advantageous to a leader when it comes to keeping in touch with what is important to the customer. There are times when a leader can lose touch of the customer desires and needs. In such case, there is a danger of not delivering what the customer is seeking. This can eventually lead to strained relationships and even a loss of client and business. Keeping in touch with the customer provides an advantage in delivering the expected value. Having the pulse of the customer allows one to discern between what the customer wants and what the customer needs. Because the leader of this project remained close to the customers and frequently sought their feedback, it was relatively easy to ensure their needs were known as changes that were being planned and implemented. The customers viewed the project as a value to them as it provided a means to have their chemical inventory tracked and significantly reduced the amount of time that a worker would potentially interrupt work activities while performing his or her inventory duties.

10.6.4 Seeking Perfection Always

A leader who seeks perfection in every aspect of his or her role as a leader is in fact on a plight to continuously improve his or her output. The leader may recognize at some point that perfection may never be achieved and that his or her actions are leading him or her to continuous improvement while in search of perfection. The project leader constantly sought out ways to increase efficiently in completing the project and increasing value for customers. The customers recognized and valued this attribute of the leader. Seeking perfection requires a strategy of continuous improvement in all associated aspects and activities.

10.7 Employee Engagement and Training

We spent a lot of time discussing the impact and importance of employee engagement in the workplace in Chapter 4. Employees were significantly engaged in identifying the issues and recommending solutions as well as ways to implement the change in getting the work accomplished. The problem identification and effectiveness of the solution devised for the project are being credited to the high level of employee engagement and creativity. The level of employee engagement was invaluable to the process and the changes that were implemented. For example, during the installation of the first batch of RFID tags, the operators noted that several tags were incomplete and if installed would create problems associated with performing the inventory in the future. One of the operators devised a method and a tool that was successfully used in screening tags to ensure completeness prior to use.

10.8 Regulatory Impact

Not all projects will impact regulatory requirements; however, an evaluation of whether there will be impacts is generally a smart option. The project presented by this case study had direct regulatory impacts. Regulatory impacts can include local, state, or federal regulations that govern the work performed by a company. For the case study, environmental and occupational safety and health regulations were applied to the task that was being performed. Therefore, considerations had to be given to the impact of these regulations on implementing the project.

The revised protocol for tracking chemicals from cradle to grave while maintaining an accurate inventory was determined by a team that included management and workers to be beneficial in ensuring regulatory compliance. Some of the realized benefits are as follows:

1. Elimination of ergonomic-related issues
2. Reduction in the risk of chemical exposure
3. Improved accuracy of chemical inventory reporting
4. Reduced hazards associated with performing the task
5. Improved process for managing hazardous waste

10.9 Environment, Health, and Safety Aspects

The primary catalysis for beginning the project stemmed from the need to address the health and safety of the workers. Under the law, workers are entitled to a safe workplace, and employers are obligated to provide a safe workplace for workers. In a safety conscious work environment, workers are encouraged to communicate environment, health, and safety concerns to their management. The issues encountered by the workers while performing the task of inventorying chemicals promptly communicated their concerns to management. Upon hearing of the issues, the management team acted promptly to protect the worker demonstrating his or her concern for the workers' well-being. This prompt action by management increased the trust that workers had in their leadership team as well as their willingness to participate in seeking solutions that eliminated the hazards associated with the task. The safety-related issues addressed with by the project include the following:

1. *Elimination of ergonomic-related issues through reducing the amount of time the workers would lend themselves to awkward body postures*: With the transition to using RFID, the need to kneel, lift, and twist frequently was essentially eliminated. Inventory was maintained

through simply walking into an area and activating the tags with a hand-held instrument.

2. *Reduction in risk of chemical exposure*: This occurred primarily because with the new process there was no need to routinely handle chemical containers to complete the inventory.

3. *Improved accuracy of chemical inventory reporting*: The new process provided increased accuracy in inventorying and tracking chemicals from the point of receiving to the point of disposal.

10.10 The People Aspects of Implementation

One important aspect of implementing Lean is not losing sight of the impact the new process improvement activities can have on the workforce. Often, the people impact is not thoroughly evaluated when implementing Lean; therefore, workers believe that they are unimportant and are not respected. In order to gain worker buy-in and support for change, people must feel that they are respected by the leadership team.

The impact of implementing a new process for inventorying and tracking chemicals from cradle to grave was destined to have impact on the workers once the project has been fully implemented. Because employee engagement was extremely high, the stress of the potential impacts to employees was minimized. The employees were actively engaged in defining their new role in the organization once the project has been implemented, recognizing the reduction in time needed to complete the task. Therefore, they were not intimidated by the project's success and felt like their management respected them and their contribution to the organization. They were vested in the project's success and worked hard to ensure that success was realized.

10.11 Case Study Implementation Pitfalls and Hidden Costs

As with implementing Lean strategy for any technology process or procedure, there is a cost associated with implementation. However, often when we hear about Lean-implementation cost, the initial cost of start-up may not be addressed. These costs can involve activities such as benchmarking, researching various aspects of the project, and planning meetings. It is feasible for the start-up cost to be more extensive than the potential value gained from implementation of a new process or procedure. Also, there is a possibility for the value of implementing Lean not being realized in the near term but in the long term.

Some of the pitfalls and hidden costs that were not originally included in the project implementation strategy for the case study are as follows:

- Cost associated with training and selection of additional staffing required to perform the conversion task. Training new workers became time consuming, and the project manager lost several trained workers shortly after training was complete for various reasons that new candidates had to be sought out and trained.
- Equipment replacement. Because of constant movement of equipment from location to location, some of the equipment failed due to the damage sustained from dropping and bumping during transit and use.
- The cost associated with developing a process to ensure batch tag accuracy and completeness and the time associated with checking each batch of tags purchased to ensure completeness.

10.12 Summary

The case study presented represents implementation of a Lean project that was successful in achieving the desired results. As discussed, there were challenges in implementation; however, those challenges were overcome through employee engagement and management support. Lean can and remains a great way for companies to improve processes and the financial bottom line. When implemented, the project's strict adherence to the project plan and strategy played a role in keeping the project on track. The performance of the project was not only monitored by management, but the workers also monitored performance and adjustments were made where necessary.

Abandoning the old process and implementing the new process formed the basis for the case study. The new process represented in the case study created efficiency in task performance, tracking chemicals from purchase to disposal, and significantly reduced the potential for adverse health effects. In addition, the new process provided a means to improve the accuracy of the chemical inventory for regulatory reporting purposes.

Because the employees were vested in the project, they were not intimidated of the potential for project success. In fact, the workers did everything within their control to ensure project was successful. The level of employee engagement served as the foundation that carried the project to the finish line.

11

Evaluation Resource Packet

The resource packet is designed to provide a means or a guide on ways to evaluate your organization when contemplating the Lean journey. It was stated in earlier chapters that before embarking upon a Lean transition, management must know with some level of assurance that the organization is ready for the journey. One way to gage organization readiness is to assess and evaluate the organization and its members. The resources in this chapter are designed to assist with this activity. Evaluation tool examples included are as follows:

1. Survey questionnaires
   a. Organization trust
   b. Organization culture
   c. Lean
   d. Organization learning self-assessment
2. Training program development guidelines
3. Training feedback forms
4. Attributes of a good assessment

The examples included represent some of the types of evaluations that may be helpful in determining if the organization is positioned for success when it comes to implementing Lean.

Organization Trust Questionnaire

Instructions: Place an X in the space that has the number that best describes your opinion of the questions being asked below. Only one number should be selected for each question, and please respond to all questions.

1—Always   2—Sometimes   3—Usually   4—Seldom   5—Never

Question#	Question	1	2	3	4	5
1.	Senior management in my company communicates information completely and frequently.					
2.	My supervisor communicates information frequently and accurately.					
3.	The managers in my organization keep promises.					

(Continued)

Question#	Question	1	2	3	4	5
4.	I feel safe while working in my workspace.					
5.	The people in my organization always follow procedures.					
6.	Clear concise communication is seen throughout the organization.					
7.	People in my organization treat each other with care and respect.					
8.	Communication flows in all directions in my company to ensure that all workers are kept informed.					
9.	Disagreements and issues are addressed in a timely manner.					
10.	My managers support a work-life balance for workers.					
11.	Managers in my organization openly admit mistakes when they occur.					
12.	Leaders in my organization are decisive decision makers.					
13.	Management always communicates openly and honestly.					
14.	Workers are actively listened to by management.					
15.	Employee benefits in my company are comparable to similar companies.					
16.	Information is freely and willingly shared by organization members.					
17.	My management team leads with confidence.					
18.	Policies are in place to ensure that employees are treated fairly across the organization.					
19.	Managers in my organization demonstrate good leadership and management skills when conducting business and making decisions.					
20.	The managers in my organization are viewed as being competent leaders.					

Organization Culture Questionnaire

Instructions: Place an X in the space that has the number that best describes your opinion of the questions being asked below. Only one number should be selected for each question and please respond to all questions.

1—Always   2—Sometimes   3—Usually   4—Seldom   5—Never

Question#	Question	1	2	3	4	5
1.	Communication flows freely in my organization.					
2.	My organization has a clear vision.					

(Continued)

Question#	Question	1	2	3	4	5
3.	Conflicts are addressed fairly and promptly.					
4.	I believe that my management values and respects my opinions.					
5.	Employees are motivated and happy to perform their jobs.					
6.	The managers in my organization celebrate success.					
7.	The managers in my organization openly acknowledge employees for a job well done.					
8.	Management seeks input from employees and are responsive to their suggestions.					
9.	The mission and values of my organization are posted, so that everyone can see them.					
10.	Employees understand their jobs and role in the organization.					
11.	The management team is trusted at all levels of the organization.					
12.	Management demonstrates a respect for employees across the organization.					
13.	Downward communication is timely and accurate.					
14.	My input is valued by my peers.					
15.	Employee speaks highly of the organization.					
16.	My supervisor is a positive role model.					
17.	Everyone takes responsibility for their actions.					
18.	Employees enjoy coming to work.					
19.	Morale is high across the organization.					
20.	Management is willing to admit mistakes openly and freely.					
21.	Everyone has the training, skills, and equipment to complete their jobs safely.					
22.	Knowledge and information sharing is common in my organization.					
23.	Management and employees are accountable for their actions.					

Employee Engagement Questionnaire

Instructions: Place an X in the space that has the number that best describes your opinion of the question being asked below. Only one number should be selected for each question, and please respond to all questions.

1—Always   2—Sometimes   3—Usually   4—Seldom   5—Never

Question#	Question	1	2	3	4	5
1.	My opinion is often sought by management.					
2.	My coworkers often ask for my input.					
3.	I willingly participate on teams to develop and implement new technology, program, or process.					
4.	I share responsibility for safety of myself and my coworkers.					
5.	I feel valued for my contribution to the success of the organization.					
6.	My immediate supervisor provides guidance on how to perform work.					
7.	I have the opportunity often to participate in providing input on decisions affecting me and my team.					
8.	I willingly provide feedback to management on issues.					
9.	Employees in my organization are motivated to accomplish work.					
10.	My management team inspires me.					
11.	I enjoy going to work.					
12.	Employees in my organization are accountable for their actions.					
13.	I am proud to tell others where I work.					
14.	I am provided the tools I need to effectively perform my task.					
15.	I know and understand how the work that I do support the goals of the company.					
16.	I am committed to doing a good job.					
17.	I would happily recommend my organization as a good place to work.					
18.	I know what my supervisor expects of me.					
19.	Employees in my organizations often proactively offer solutions to problems to management when issues are discovered.					
20.	I am satisfied with my job.					
21.	Management encourages worker involvement in organizational activities.					

Lean Questionnaire

Instructions: Place an X in the space that has the number that best describes your opinion of the questions being asked below. Only one number should be selected for each question, and please respond to all questions.

1—Strongly agree 2—Agree 3—Disagree 4—Strongly disagree
5—Unknown 6—Not applicable

Question#	Question	1	2	3	4	5	6
1.	My organization implemented Lean to reduce headcount.						
2.	My organization implemented Lean to improve performance.						
3.	My organization implemented Lean to improve financial performance.						
4.	My organization uses Lean tools and techniques.						
5.	The people in my organization think Lean is valuable to improving the organization.						
6.	Management communicates the reason for Lean implementation clearly.						
7.	Employees are resistant to change.						
8.	Management is resistant to change.						
9.	Employees were trained in the process after Lean was implemented.						
10.	After Lean was implemented, the process was improved.						
11.	A plan to implement Lean was developed.						
12.	Waste is often identified and removed from the system.						
13.	The Lean project plan was effective.						
14.	Do you think implementation of Lean was effective?						
15.	Management keeps in touch with the customers and knows what they expect.						
16.	Management shares with the worker the expectation and needs of the customer.						
17.	Did management prepare workers for the change associated with Lean implementation?						

Organization Learning Self-Assessment

Instructions: Place an X in the space that has the number that best describes your opinion of the questions being asked below. Only one number should be selected for each question, and please respond to all questions.

1—Strongly agree   2—Agree   3—Disagree   4—Strongly disagree
5—Unknown

Question#	Question	1	2	3	4	5
1.	Training programs are designed to help individuals close the gap in learning and achieving their learning goals.					
2.	This organization is a learning organization. We learn from our mistakes.					
3.	This organization is constantly learning from past performance to improve performance.					
4.	Management encourages learning from success and failures.					
5.	Lessons learned are developed and distributed across organizational lines for purpose of sharing information and learning.					
6.	Information is shared fluidly among teams.					
7.	Frequent feedback is provided to individuals on job performance.					
8.	It is easy to discuss with management what is needed to improve his or her performance.					
9.	Organization members share knowledge freely.					
10.	The organization solicits feedback from customers and stakeholders for purpose of developing and improving training.					
11.	Teams are frequently developing more effective ways of performing group tasks.					
12.	Training programs are developed to help team members achieve their learning goals.					
13.	The organization is open to learning through various means and sources.					
14.	Managers, coaches, and mentors help individuals develop and implement learning objectives.					
15.	Employees are encouraged to continue to enhance their knowledge.					
16.	Management provides funds to ensure employees receive the needed training.					
17.	Employees embrace learning as a means to help the organization achieve success.					
18.	Experimentation that supports learning is supported by management.					
19.	Requesting feedback from employees is a constant effort in my organization.					

Training Program Development Guidelines

I. Identify the goal of the training

It is difficult to develop a successful training program without knowing exactly what you are trying to achieve. The goals that are set for the training will determine the approach to program development. A list of questions that can be used to help flush out and document your training goals is as follows:

1. Is there a need to improve workers' knowledge level as a means to improve performance?
2. Is training needed to improve workers' performance in current job?
3. Is there a need to prepare workers for newly developed positions or modified positions?
4. Is training needed to prepare employees for promotion?
5. Is training mandated by a procedure or regulation?
6. Is training required to reduce accidents, injuries, and illness, and to improve safety practices?
7. Is the training required to improve product quality or productivity?
8. Is the training required to orient new employees to the job?
9. Is training required to teach employees how to train other employees?

II. Analyze the training needs

Analysis of training needs should address the following:

1. What the organization is seeking to accomplish?
2. What needs will training address?
3. What are the process and procedural changes?
4. Who needs to be trained?
5. What are the procedure and process changes?
6. What resources are available to conduct training?
7. Identify a subject matter expert (SME) for the training topic and content.
8. What knowledge and skills are needed?
9. What are the roles and responsibilities?

III. Design the training program

Before training program can be developed, a decision must be made on the type of training needed. One type of training method may not be effective in achieving all of the objectives or goals. Therefore, it is important to review the advantages of each type in relationship to your objective or goal. Below are some questions that can be asked to assist in determining the type or types of training method that should be considered in development.

1. Is it feasible to train on the job?
2. Should training be conducted in a classroom setting?
3. Will a combination of scheduled on-the-job training and traditional classroom instruction work best?
4. Can the classroom training be completed in one session or will multiple sessions be required?

IV. Develop the training program
Acquire training resources: Training materials may need to be purchased from a vendor, or there may be a need to involve several vendors. This might include computer software, an online course, or books. In some cases, there may be a need to customize certain parts of the material, so that the materials address the training objectives. In some cases, the material must be developed.

Identify trainer: Recruit an experienced professional to deliver and guide employees through the training. It may be beneficial to use an in-house trainer who is experienced in your business, or you may need to hire a professional trainer and brief him or her on the important aspects of the business. The trainer will lead trainees, give lectures, answer questions, and provide feedback.

V. Implement the training program
Conduct training: Ensure the trainer is prepared and qualified to deliver the material. While conducting the training, consider ways to capitalize on the following:

1. Employee participation.
2. Use humor when feasible. Do not overdo it with the humor because you may lose some of the audience.
3. Package the lesson in a way that it is easy to maneuver through and find key information while they are being discussed.
4. Make the learning opportunity fun. If the trainee finds the material fun, his or her retention becomes enhanced.

VI. Feedback mechanism

Upon completion of each training class, it is a good idea to obtain feedback from participants. This is a great opportunity to gain information on the strengths and weaknesses of the training program, what was learned, and the participant's overall experience. Feedback can be obtained through interviews or requesting attendees to complete a survey immediately after course completion. This information can help to improve the program for future classes.

VII. Evaluate the training program

The purpose of evaluating the training program is to ensure that it is effective and is successful in filling the gap or providing the required knowledge for employees. When designing an evaluation process, one must ensure that the following four concepts are addressed:

1. *Reaction*: Can include the learning environment, instructor, or learning material.
2. *Learning*: Did the participant learn the material?
3. *Transfer*: Did the trainee behave differently when he or she returned to his or her jobs?
4. *Results*: Did the training have no impact on business outcome?

Questions that should be answered during the evaluation process include the following:

1. How many people were trained?
2. Was the objective of the training met? If not, to what extent?
3. Was training administered in the appropriate facilities with the appropriate materials?
4. What improvements were made to the program over the past 12 months?
5. Was feedback solicited from participants?
6. Was feedback incorporated in the training to improve delivery and content?
7. Was the cost of training impacted as a result of improvement? If yes, to what extent?

Training Feedback Form

Date: _____

Title of Training: _____

Location: _____

Instructor/Trainer: _____

Instructions: Indicate your level of agreement with the statements provided regarding the training by placing an "X" in the appropriate box alongside the statement.

	Strongly Agree	Agree	Neutral	Disagree	Strongly Disagree
The objective of the training was clearly defined and stated					
The content was well organized and easy to follow					
The topics covered were relevant to me in performing my job					
Participation and engagement were encouraged and welcomed					
This training is useful and will help me perform my job better					
The trainer was knowledgeable in the material delivered					
The trainer was well prepared to teach the class					
The class material distributed were useful					
The training objectives were met					
Appropriate time was allotted for the training					
The training environment was comfortable					
The length of the course was appropriate					

Training Feedback Form

Date: _____

Title of Training: _____

Location: _____

Instructor/Trainer: _____

Please answer the following questions:

1. What did you like most about the training?
2. What did you like least about the training?
3. How will you use this training in performing your job?
4. What aspects of the training would you like to see improve?
5. Would you recommend this training to other workers? Why or why not?
6. Please use the space below to provide additional feedback.

Attributes of a Good Assessment

An assessment report should be written in a way that it is easily understood by the reader, and the content is relevant to the topic and purpose of the assessment. When writing an assessment report, consider the following as a starting point to organize your thoughts and the report.

1. The purpose of the assessment
   There are many reasons why conducting an assessment is beneficial to a program and ultimately the company. However, three main purposes of program assessment are:

   - *To improve*: The assessment process should provide feedback to determine where and how programs, processes, practices, and procedures can be improved.
   - *To inform*: The assessment process should inform management of the status and functionality of the program or process being evaluated.
   - *To prove*: The assessment process and results should demonstrate to workers, the leadership team, regulatory community, and other outside interested parties how the program and subsequently the company are performing in the area or areas assessed.

2. Selection of the assessment team where feasible
   When selecting the assessment team, ensure that the team members have the appropriate skills and credential to provide valuable input on the way the process is functioning as well

as able to recommend viable improvement actions that can be implemented. When writing the assessment report, ensure to include a brief credential summary for each team member.

3. Methodology used to collect data

This section should carefully outline the process used to complete the assessment and data collection. Data collection methods may vary depending on the business and the processes being assessed. Data collection can include the following:

- Document reviews (policies, procedures, and so on)
- Interview of workers, management, or customers
- Previous assessment and audit reports (ones performed internally as well as externally)
- Metrics
- Incidence reports
- Injury and illness reports
- Task or work observations

4. Conclusions

The conclusions should contain what was found during the assessment that support current business practices and outcomes that are noteworthy or issues that must be addressed to improve business outcomes.

5. Recommendations

Recommendations should be clear, concise, and actionable. Refrain from making individual judgment calls that support one's preconceived judgment and experiences.

6. References and attachments

In this section of the report, include items and activities such as

- A list of document reviewed.
- A list of personnel interviewed by title (never include names of interviewees).
- Credentials for the assessment team members.

Index

Note: Page numbers followed by f and t refer to figures and tables, respectively.

Milton Keynes UK
Ingram Content Group UK Ltd.
UKHW031150141024
449569UK00024B/919